Mathematics for Chemistry

Graham Doggett
and
Brian T. Sutcliffe

Essential Maths for Students

Mathematics for Chemistry

Graham Doggett
and
Brian T. Sutcliffe

Chemistry Department,
University of York, Heslington,
York, YO1 5DD

Longman
Scientific &
Technical

Longman Scientific & Technical
Longman Group Limited
Longman House, Burnt Mill, Harlow
Essex CM20 2JE, England
and Associated Companies throughout the world

Copublished in the United States with
John Wiley & Sons, Inc., 605 Third Avenue, New York
NY 10158

Longman Group Limited 1995

First published 1995

British Library Cataloguing in Publication Data
A catalogue entry for this title is available from the British Library.

ISBN 0-582-21970-1

Library of Congress Cataloging-in-Publication data
Doggett, Graham.
    Mathematics for chemistry / Graham Doggett and Brian T. Sutcliffe.
      p.   cm.
    Includes bibliographical references and index.
    ISBN 0-470-23483-0
    1. Mathematics.   I. Sutcliffe, Brian T.   II. Title.
QA37.2.D64   1995
515′.1′024541–dc20   94–39285
                                     CIP

Typeset by 21WW in 10/12pt Times New Roman.

Contents

Introduction

Goldberg, 1988

Mathematics in the context of chemistry

Chemistry is a practical subject, so why should mathematics now play such an important role in its understanding? Coulson provided compelling answers to this question in his presidential address to the Institute of Mathematics and its Applications (Coulson, 1973), when he reviewed the reactions of those involved in the development of chemical ideas one hundred years earlier. He reminds us, for example, that in 1878 Frankland wrote: 'I am convinced that further progress of chemistry as an exact science depends very much indeed upon the alliance with mathematics'. This prophetic view was not shared by most chemists of the time; and it was not until the development of the quantum theory in the late 1920s, and the consequent impact on our understanding of spectroscopy and electronic structure, that chemists started to develop the mathematical tools that were relevant to the needs of chemistry. There are many reasons for the growth of this symbiotic relationship, and it is helpful to examine some of them in putting the objectives of this book into the proper context.

The study of chemistry, whilst concerned intimately with the syntheses and reactions of an ever increasing number of compounds, is concerned basically with the discovery of patterns in the way chemical properties of such compounds are interrelated. At the simplest level, for example, the shell model of the atom (a mathematical concept) relates to the framework provided by the Mendeleyev classification of the elements – the detailed understanding of which requires mathematical insight to see how the periodic classification is manifested in the quantum mechanical concept of the orbital model. The point about this development is that the orbital model in turn provides an excellent tool for understanding the nature of atoms and molecules, the microscopic behaviour of which may then be explored with the aid of experimental spectroscopic techniques.

The intrusion of mathematics into chemistry provides the necessary tools for quantitative model building that are required for the prediction, elucidation and rationalization of chemical phenomena. It is very difficult indeed, for example, to recognize and verify the presence of both simple (Fe^{16+}) ▷

▷
Taylor and Williams,
1993

and unusual chemical species like $HC_{3+2n}N$ ($n = 0, 1, 2, 3, 4$) in interstellar space ▷, without the orbital model for atoms and molecules. Furthermore, without the underlying warp of mathematics, and the weft of physics and biology, chemistry would be reduced to a vast catalogue of apparently unrelated facts and observations: instead of quantitative models there would be 'rules of thumb'! It is because of this strong interrelation between mathematics and chemistry (and physics) that we are able to understand the molecular structure of biomolecules such as insulin, through the interpretation of the results obtained from X-ray diffraction. Understanding molecular structure is, of course, a precursor to understanding the chemical behaviour of atoms and molecules.

Other branches of chemistry, whilst less concerned with the determination of molecular structures, are concerned more with the gathering of data by observing chemical species reacting: in these situations, where time is the key variable, the results can only be interpreted and understood with the aid of a knowledge of the form and solution of special kinds of differential equation. The chemical objective here is to interpret the observed results in terms of a mechanism for the reaction, and this necessitates plotting data in the form suggested by the theory in order to recognize the function that relates concentration of a species and time. However, because there are errors present in the data collected, there are problems associated with the handling of these errors when attempting to establish the quantitative relation between concentration and time. The proper treatment of the problems experienced in these kinds of situation involves understanding the ideas of error propagation and statistics – both of which involve using the tools of calculus. In fact, estimating and assessing the consequences of error propagation in some form or other pervades all experimental science.

This book is written to help those for whom mathematics has always been a problem, because the subject has not been studied to the depth required for understanding the infrastructure of chemistry. We hope that it will also help those who have studied mathematics in more depth, but the interrelation between their knowledge and the requirements of chemistry is not fully developed. Typically, although this latter group may have an elementary knowledge of complex numbers and even group theory, the connections with structural, spectroscopic and quantum chemistry are unlikely to have been made. Furthermore, despite their knowledge of mathematics before entering higher education, much of the detail of what has been learnt earlier has become forgotten and unused. In fact, there is often a feeling of anxiety (and associated lack of confidence) in many students when it becomes apparent that important areas of chemistry are going to become inaccessible without some sound working mathematical tools in their baggage.

In view of these well-known problems, we have set ourselves the aim of raising the threshold for the onset of anxiety, by presenting a selection of mathematical ideas and techniques in the form of a tool-kit set within a chemical context. We therefore adopt the deliberate policy of avoiding much formal proof in favour of a more pragmatic approach – simply because the mathematics of chemistry is linked to physical phenomena and, for the most part, the exception to the rule is not the norm.

The material covered in this text splits broadly into the two areas of calculus and linear algebra, prefaced by an introductory tour through some necessary mathematical grammar and symbolism. The fact that the models used in understanding and rationalizing real physical problems can involve the simultaneous use of tools from each of these areas (and also the deployment of other techniques) is recognized from the outset in our choice of topics and is seen, for example, in the ubiquitous and apparently simplest of problems in requiring the best fit of observed data to a linear or quadratic form.

Organization of the text

The text is organized so that the first three chapters provide a review of elementary principles and notation in algebra, trigonometry and calculus. This review can be regarded as revision or a rapid survival course, depending upon the background of the reader! There is then a reiteration of some of the calculus before further developments and applications are considered within the general contexts of kinetics, thermodynamics and spectroscopy. The remainder of the text is concerned more with topics that come within the area of mathematics termed linear algebra, where time is spent developing techniques involving the matrix and vector notation, in preparation for all kinds of applications (primarily those associated with spectroscopy and bonding theory). The inclusion of these topics is most important as they also provide the tools for dealing with the anisotropy of directional chemical properties associated with, for example, the consequences of electric and magnetic fields interacting with matter. We are quite honest about our general style of approach and choice of content. In no way can we hope to be exhaustive, and we make this clear at the outset. Our basic aim is to provide some useful tools which, in many cases, act like keys for opening doors into other areas of mathematics. Thus, for example, armed with ideas of functions, calculus and complex numbers, it is possible to gain access to the theory of functions of a complex variable as a first step in developing the understanding of the theoretical modelling of scattering processes and advanced spectroscopic methods used in the laboratory – just to highlight two applications.

Throughout the text marginal notes are used for comment, and citations of references, equations, sections, etc. All references cited by author name are collected in the References section at the end of the text. Answers are also included for all the problems given in the text, along with some working and hints. Examples and problems of a chemical nature are used as far as possible, in the knowledge that the detailed chemistry will require students to consult a physical chemistry text. It is almost inevitable that some of the examples will be premature; however, the problems are self-contained and, for their execution, do not rely on a detailed knowledge of the chemistry. In developing the mathematics, we use current widely accepted physical chemistry texts either to illustrate the application of mathematical principles, or to provide additional mathematical commentary. Further examples and problems in the more basic aspects of mathematics may be found in *Foundation Maths* by Croft and Davison.

Acknowledgements

We would like to express our appreciation to several groups of people who have helped us in the preparation of the manuscript: the reviewers for their positive comments, suggestions and encouragement; Tamsin Mansley for her critical reading of the first draft; Robert Gammon and Peter Lee for their valuable comments on individual chapters; Alastair Doggett for his help in preparing the diagrams using UNIRAS; Robert Fletcher for advice on the use of UNIRAS, and Stephen Smith for help with the finer aspects of LaTeX.

BTS is much indebted to the Quantum Theory Project at the University of Florida for hospitality, during which the manuscript was being completed.

Finally, we thank Kathy Hick and her colleagues in the Longman Editorial Office for their patience, constant help and encouragement. All errors are, of course, our responsibility!

Graham Doggett
Brian Sutcliffe
May 1994

1 Numbers, symbols and rules

Objectives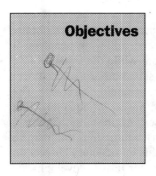

This chapter provides

- a working knowledge of some of the very basic building blocks associated with numbers, and the rules determining their manipulation
- confidence in handling numbers with associated units
- a simple introduction to algebra and equations
- an insight into the need for extending the number system to include complex numbers

Mathematics, like much of chemistry, is concerned with numbers, symbols, and rules for their manipulation. Searching for pattern also forms an important part of the rule development. The complexity of the symbolism in mathematics, delineated by particular rules, often obscures meaning and comprehension as, within a given subject area, the background knowledge required is often hierarchical in nature, and the consequent understanding of the working of the mathematics may become difficult – especially if there are gaps in this background knowledge.

The development of ideas in chemistry follows a very similar hierarchical pattern and much of the associated thinking and experience is basically related to working with numbers, symbols and rules. Numbers permeate chemistry in terms of experimentally measured values, or in the appearance of quantum numbers, associated with physical properties. Although symbols are used in chemical equations, according to well-defined rules, several layers of meaning are usually associated with such equations which are, in effect, mathematical statements. In many cases it is the greater familiarity with chemical symbolism that makes it feel more comfortable; however, in many situations, it is the implicit message carried by the symbolism that is important. In the mathematical situation it is therefore more difficult for the chemist to perceive, or even appreciate, the significance of what is not immediately apparent. An example will make this clear: when dealing with numbers or symbols in mathematics or chemistry, the context makes it clear whether order (*commutativity*) is important. Thus $2 \times 3 = 3 \times 2$ or $xy = yx$, if x and y are symbols representing numbers multiplied together (it is common practice to omit the × sign when there is no ambiguity). The basic rules of handling numbers include commutativity for + and ×, which is nearly always assumed. In chemistry, the analogous symbolism has to be read in a different way, because IBr is not the same as IrB (or BIr!). Given the symbols for the elements there are special rules that inform the chemist about how to order these symbols and manipulate them in equations: thus HCN and HNC are recognized as isomers, but for $FeSO_4$ and $FSeO_4$, the latter compound has no sense when interpreted in terms of the rules of valence. Thus, it is clear that in any symbolic approach of classification or definition, the rules for using the symbols must be recognized; also, there may be other factors that need to be understood – both in the mathematical and chemical situations. Many examples of the former will be discussed in this and following chapters; it merely suffices to say that, in the chemical situation, while the equation $C + 2H_2O = CH_4 + O_2$ may make sense in term of conservation of matter, it may not make sense in terms of either kinetics or thermodynamics (or both!). The chemist is aware of the background to the use and misuse of equations in a chemical context; when it comes to the same sort of ideas set out in a mathematical context, it is the background (implicit) assumptions that may be absent or undeveloped. We hope, therefore, that in this introductory chapter on numbers and symbolism, we may prepare the reader for further developments and explorations of areas of mathematics that provide important tools for the chemist. The first section of this chapter is concerned with numbers of different kinds, and the rules for their manipulation; this is followed by sections on algebra and equations before returning to a discussion of an extension of the number system to include complex numbers.

1.1   Numbers

Kinds of numbers

Real numbers come in various kinds: integer, rational, decimal, irrational. *Integers* are the counting numbers (no decimal point) extended to include zero and negative values: thus the *set* $\mathbb{I} = \{\ldots, -1, 0, 1, \ldots\}$ contains all the integers. Integers can be either *odd* or *even*, depending whether or not they are divisible by 2; an integer p, not divisible by another integer (apart from ± 1, $\pm p$) is called a *prime* number. *Rational* numbers, or fractions, are of the form $\frac{r}{s}$ or r/s (for example 2/3) ▷, where r, s are integers ($s \neq 0$) corresponding to the *numerator* and *denominator*, respectively. Clearly, if $s = 1$, we see that integers are included in the set of rational numbers. A rational number can always be written as a *decimal* number, though not necessarily in terminating form: for example $1/3 = 0.333\ 33\ \cdots$. Decimal numbers which have an infinite number of digits after the decimal point, but with no repeating pattern,

▷

The symbol / for division is sometimes referred to as a *solidus*.

▷
See Chapter 2.

are termed *irrational* numbers, and some such numbers play an important role in chemistry: for example, π, $\sqrt{2}$, e (the base for natural logarithms ▷), etc.

A decimal number that can be written as a rational number must possess a finite number of digits after the decimal point or an infinite number with a repeating pattern. Thus, for example, 1.128 can be written as a rational number, since $1.128 = 1.128 \times 1000/1000 = 1128/1000 = 141/125$, after cancelling the common factor of 8 from numerator and denominator. On the other hand, for the number with the repeating pattern of 128, we can proceed as follows: let $x = 1.128\,128\cdots\cdots$, then $1000x - x = 1127$ implies ▷ $x = 1127/999$.

▷
The symbol $\Rightarrow$ is usually used for *implies*.

Problem 1.1

Write the following decimal numbers in rational number form: $0.005\,5$, $0.055\,555\cdots$, $0.037\,037\cdots$.

The necessity for recognizing the difference between an integer and a decimal number is important within the context of computing, as a decimal number could take twice the storage space of an integer. Thus, a decimal number essentially requires storage locations for the integers before and after the decimal point (but this is not exactly how the number is stored in practice). The storage of irrational numbers presents an especially awkward problem because they must always be truncated to a decimal number with a fixed number of digits after the decimal point: the computer software (and hardware) therefore determines the size of the error in storing such numbers. This problem can be demonstrated on most electronic calculators by displaying the number 2 and then pressing the square root button six times (say). If the squaring button is now pressed six times, then a decimal number close to 2 is usually obtained.

Relations involving numbers

▷
Known as *inequalities*.

▷
Thus multiplying both sides of an inequality by a negative number changes the sense of the inequality.

Apart from magnitude and sign, real numbers can be ordered in an increasing or decreasing sense: so, for example, -1 is less than 2, but 2 is greater than -1; also -1 is greater than -5, etc. These ordering relations ▷ are written symbolically as $-1 < 2, 2 > -1, -1 > -5$, respectively. In general terms x is larger than y if $x - y$ is a positive number; if $x < y$ then $x - y$ is a negative number. Thus $-2 < -1$, but $2 > 1$ ▷. If x is very much larger than y, then we often write this relation as $x \gg y$.

From a mathematical point of view it is often necessary to discuss collections of numbers, rather than focusing attention on individual numbers. In this sort of situation, which we shall meet first in the discussion of functions, we use the shorthand $\mathbb{R}$ to represent the whole collection (set) of

$\triangleright$
For an individual integer, j, we write this as $j \in \mathbb{I}$ or $j \in \mathbb{R}$, depending upon the context.

real numbers. From time to time it will be necessary to use other shorthand forms to represent parts (*subsets*) of this complete collection. So, for example, the set of all integers, $\mathbb{I}$, is contained in $\mathbb{R}$, and we write this as $\mathbb{I} \subset \mathbb{R}$ $\triangleright$; similarly, we may wish to refer to the subset of all positive numbers.

Operations on numbers

The manipulation of expressions containing fractions involves the use of a few simple rules:

Addition or subtraction

Here all fractions are rewritten with a lowest common denominator (the lowest number into which all denominators will divide).

Worked example

1.1   (a) $\dfrac{1}{6} + \dfrac{1}{2} - \dfrac{1}{3} = \dfrac{1}{6} + \dfrac{3}{6} - \dfrac{2}{6} = \dfrac{1+3-2}{6} = \dfrac{2}{6} = \dfrac{1}{3}$.

With practice, it is possible to miss out the first step involving the rewriting of the (three) fractions in terms of the common denominator.

(b) $\dfrac{2}{3} - \dfrac{1}{9} + \dfrac{1}{2} = \dfrac{12 - 2 + 9}{18} = \dfrac{19}{18}$, where 18 is the lowest common denominator.

(c) $\dfrac{2}{x} + \dfrac{1}{3y} = \dfrac{6y + x}{3xy}$, where x and y are symbols representing arbitrary numbers, neither of which is zero.

Multiplication

$\triangleright$
Indicated by $\cdot$ or $\times$ (the latter when there is any ambiguity with the sign for the decimal point).

Multiplication of fractions follows the simple rule of taking the product $\triangleright$ of the two numerators divided by the product of the two denominators.

Worked example

1.2   (a) $\dfrac{2}{3} \cdot \dfrac{1}{6} = \dfrac{2}{18} = \dfrac{1}{9}$, after cancelling the common factor 2 in both numerator and denominator.

(b) $\dfrac{a}{b} \cdot \dfrac{c}{d} = \dfrac{ac}{bd}$.

(c) $\dfrac{1}{3} \cdot \dfrac{c}{d} = \dfrac{c}{3d}$.

Division

Division of a number by a fraction r/s is accomplished by multiplying the given number by s/r.

Worked example

1.3   (a) $\dfrac{2}{3} \div \dfrac{1}{2} = \dfrac{2}{3} \cdot \dfrac{2}{1} = \dfrac{4}{3}$.

(b) $\dfrac{a+b}{2} \div \dfrac{x}{4} = \dfrac{(a+b)}{2} \cdot \dfrac{4}{x} = \dfrac{2(a+b)}{x} = 2(a+b)/x$.

▷
Without the use of brackets $a + b/x$ is interpreted as the addition of a to b/x.

Brackets are used to ensure that addition of a and b is carried out prior to division by x ▷. Notice that if the solidus is used instead of $\div$, then

$$\frac{(a+b)}{2} \bigg/ \frac{x}{4} = \frac{(a+b)}{2} \cdot \frac{4}{x} = 2(a+b)/x.$$

Problem 1.2

Simplify:

(a) $5/10 \div 1/2$, $\ 2 \div 4/3$, $\ x/y \div 4x/3y$;

(b) $7(5/2)(6/5)$, $\ (5/2)(3/4) \div 3$, $\ 1 + (5/2)(7/6) + (3/4)/3$.

Exponential notation

Consider the number $\frac{2}{3}$ multiplied by itself four times:

$$\frac{2}{3} \cdot \frac{2}{3} \cdot \frac{2}{3} \cdot \frac{2}{3} = \frac{16}{81} = \frac{2^4}{3^4}.$$

The number 16 in the numerator is written as 2^4, since 2 is multiplied by itself four times; similarly, 81 in the denominator is written as 3^4. Numbers written with an index in this manner are said to be given in exponential form, where 2 (or 3) is the base and 4 the *exponent* or *index*.

Laws of exponents

▷
Written symbolically as $x \neq 0$.

For a given non-zero number x ▷, the exponents obey the following laws:

(1) $x^0 = 1$.

(2) $x^n x^m = x^{n+m}$. From law (1) we can write $1 = x^0 = x^{(m-m)} = x^m \cdot x^{-m}$; hence dividing both sides of this equality by x^m yields $1/x^m = x^{-m}$.

▷
See Example 1.3.

(3) $x^n/x^m = x^n \cdot 1/x^m$ (from the division rule for fractions ▷) $= x^{n-m}$ from the above deduction from law (2).

(4) $(x^n)^m = x^{nm}$, since the notation means x^n multiplied by itself m times, and the result is obtained directly as a consequence of law (2).

(5) $(xy)^n = x^n \cdot y^n$; again this result follows from law (2) together with the commutative property of multiplication on ordinary numbers (order of multiplication is not important).

Worked example

1.4 (a) $2^3 \cdot 2^4 = 2^7$ (b) $(2^3)^2 = 2^3 \times 2^3 = 2^{3 \times 2} = 2^6$

(c) $2^3/2^4 = 2^3 \times \dfrac{1}{2^4} = 2^{3-4} = 2^{-1} = \dfrac{1}{2}$

(d) $\dfrac{8}{27} = \dfrac{2^3}{3^3} = \dfrac{2}{3} \cdot \dfrac{2}{3} \cdot \dfrac{2}{3} = \left(\dfrac{2}{3}\right)^3$ or

$\dfrac{8}{27} = \dfrac{2^3}{3^3} = 2^3 \times 3^{-3} = (2 \times 3^{-1})^3 = \left(\dfrac{2}{3}\right)^3$

(e) $(3x^2)^2 = (3x^2)(3x^2) = 3^2 x^4 = 9x^4$,   (f) $(1+x^2)^{-1} = \dfrac{1}{(1+x^2)}$.

Problem 1.3

Simplify:

$$(3/2)^{-2}/(4/5), \quad (2^{-2} - 3^{-2})^{-1} \cdot (2 \times 3)^{-1}, \quad (1 + 2^{-1} + 2^{-2} + 2^{-3}),$$
$$(1 + 2^{-1})(1 - 2^{-1})^{-1}.$$

▷
The mth root of a number y is a number such that its mth power is x, i.e., $y^m = x$. $x^{1/2}$ is a special case and is written as $\sqrt{x}$.

Rational exponents

The laws of exponents (1)–(5) apply to rational exponents as well:

(6) $x^{1/m}$ means the mth root of x ▷, sometimes written as $y = \sqrt[m]{x}$ (there could be more than one value for y but, for all such y, $y^m = x$).

(7) $x^{n/m}$ means the mth root of x^n or the nth power of $x^{1/m}$.

(8) $(xy)^{n/m} = (x^n y^n)^{1/m} = x^{n/m} y^{n/m}$.

Real number exponents

The laws $(1) - (8)$ for integer and rational exponents apply for all real numbers, including irrationals.

Worked example

1.5   (a) $4^{-\frac{3}{2}} = \dfrac{1}{4^{\frac{3}{2}}} = \dfrac{1}{\left(4^{\frac{1}{2}}\right)^3} = \dfrac{1}{2^3} = \dfrac{1}{8}$. Note that $-\frac{1}{8}$ is another solution, as -2 is also a square root of 4.

(b) $(2x^2)^{\frac{1}{2}} = 2^{\frac{1}{2}}x$ (from law (5)).

(c) $(xy^3)^{\frac{2}{3}} = x^{\frac{2}{3}}(y^3)^{\frac{2}{3}}$ (from law (8)) $= x^{\frac{2}{3}}y^2$ (from law (4)).

▷
Atkins, p. 37.

▷
Note that the units of molar mass are kg mol^{-1}.

▷
Alberty and Silbey, p. 599.

Problem 1.4

(a) Simplify:

$$\sqrt[3]{\frac{2}{3}\cdot\frac{2}{3}\cdot\frac{2}{3}}, \quad (\sqrt[3]{x})^4, \quad 25^{-\frac{3}{2}}, \quad 27^{\frac{2}{3}}\times 9^{\frac{1}{3}}, \quad ((a^{\frac{1}{2}})^2/b^{-\frac{3}{4}})/(ab)^{-\frac{1}{3}},$$
$$\sqrt{4mx^4y^3}, \quad \sqrt{(8xy^2)(2x)}.$$

(b) Atkins ▷ demonstrates that, in the penultimate step in calculating the mean speed, $\bar{c}$, of a gaseous atomic or molecular species of mass m, and at a temperature T, the expression

$$\bar{c} = 4\pi\left(\frac{M}{2\pi RT}\right)^{\frac{3}{2}}\times\frac{1}{2}\left(\frac{2RT}{M}\right)^2$$

is obtained, where R is the gas constant and M the molar mass ▷ of the atom or molecule. Show that

$$\bar{c} = \left(\frac{8RT}{\pi M}\right)^{\frac{1}{2}},$$

and calculate the value of $\bar{c}$ for N_2 molecules in air at 298 K.

(c) Alberty and Silbey ▷, in the section of their textbook on the kinetic theory of gases, derive the average kinetic energy of a molecule in the x-direction, $\overline{v_x^2}$, in the form

$$\overline{v_x^2} = \left(\frac{m}{2\pi kT}\right)^{\frac{1}{2}}\frac{\pi^{\frac{1}{2}}2}{2(m/2kT)^{\frac{3}{2}}2}.$$

Show that $\overline{v_x^2} = kT/m$.

Scientific notation

For numbers with many digits, a space is placed after every three digits when there are more than four digits before or after a decimal point.

Chemistry, just like astronomy, is concerned with very very small and very very large numbers. In order to handle these kinds of numbers, it is helpful to use the so-called scientific notation in which numbers like $\pm 1\,000\,000.0$ are expressed in the form $\pm a \times 10^p$, where a is greater than or equal to 1.0 and less than 10.0, and p is an integer. So, for example,

$$-1\,000\,000.0 = -1.0 \times 10^6 \text{ and } 126.546 = 1.265\,46 \times 10^2,$$

and observe that the power of ten (p) is just the number of moves of the decimal point to the left we make in order to find a if the number is greater or equal to 10.0. To find a for a number less than 1.0, we move the decimal point to the right in an analogous manner:

$$0.000\,012\,34 = 1.234 \times 10^{-5} \text{ and } 0.918 = 9.18 \times 10^{-1}.$$

Notice that, in both examples, p is now a negative integer.

Numbers in chemistry define values of physical or chemical properties; however, such properties usually carry units. In the SI system of units, all physical properties are expressed in terms of the basic units of mass (kilogram, kg), length (metre, m), time (second, s), electric current (ampere, A), thermodynamic temperature (kelvin, K), and amount of substance (mole, mol). A useful extended compilation of physical quantities and their units can be found in several published sources $\triangleright$.

$\triangleright$
See, for example, Mills *et al*.

As already noted above, numbers can either be very large or very small: thus, for example, there are $6.022\,14 \times 10^{23}$ atoms in 0.012 kg of ^{12}C, yet the binding energy of the outermost electron in one carbon atom is $1.803\,4 \times 10^{-18}$ J.

The laws of exponents can be used when manipulating numbers in the scientific notation:

$$1.234 \times 10^{-5} \times 3.7 \times 10^3 = 3.7 \times 1.234 \times 10^{-2} = 4.565\,8 \times 10^{-2}.$$

Notice that the full number of digits possible after the decimal point has been given because all the numbers are of significance; but if 3.7×10^3 is obtained from an experimental measurement, then we have only *two* significant figures. In these circumstances we can only expect two significant figures at best in the product, and the answer is given as 4.6×10^{-2}.

Comment

$\triangleright$
See Appendix 1.

Regular usage of properties with a value given in negative or positive powers of ten is usually made simpler by introducing multiples $\triangleright$ of the basic property unit: thus, for example a volume of 10^{-3} m^3 = $(10^{-1}$ m$)$ $(10^{-1}$ m$)(10^{-1}$m$)$ = $(10^{-1}$ m$)^3$ = 1 dm^3 ; a bond energy of 4000 J mol^{-1} = 4×10^3 J mol^{-1} = 4 kJ mol^{-1}; the ionization energy for the hydrogen atom is 2.1797×10^{-18} J = 2.1797 aJ.

Problem 1.5

▷
That is, $1.0 \leqslant a < 10.0$.

(a) Write each of the following expressions in the form $a \times 10^p$, where a lies between 1 and 10 ▷:

$$(1.23 \times 10^4)(0.056 \times 10^{-2}), \qquad \frac{1.2 \times 10^{-4}}{0.6 \times 10^2},$$

$$1.2 \times 10^{-4} + 0.60 \times 10^{-2}.$$

▷
See p. 3, for example, where the numerical values of the temperatures are listed under a column with heading T/K and, for $T/K = 216.55$, we deduce that $T = 216.55$ K

▷
See Appendix 1.

(b) In Mills *et al.* ▷ it is recommended that tabulated values of physical properties should be presented with a column heading containing the symbol for the property together with the units, such that the column entries are pure numbers. For the numerical value of the temperature of 4.6179 listed under the column headed by 10^3 K/T, give the value of $1/T$.

(c) The specific heat capacity at constant pressure, c_p, of platinum is listed in a table of physical properties under a column headed $10^{-2}c_p$/J kg^{-1} K^{-1} as 1.4; give the value of c_p in kJ kg^{-1} K^{-1}.

(d) If a carbon–carbon bond length is listed as 0.140 nm, give its value in pm ▷.

(e) The stretching frequency of the carbonyl bond in methanal is at 1760 cm^{-1}. What is this value in m^{-1}?

Problem 1.6

(a) Given that the density of benzene (molar mass 0.0781 kg mol^{-1}) is 879 kg m^{-3}, and that one molecule has an effective surface area of 2.5×10^{-19} m^2, estimate the area of a monomolecular layer that can be produced from 1 cm^3 of benzene.

(b) Calculate the energy, E, of the electron in the ground state of the hydrogen atom using the equation

$$E = -\frac{me^4}{8h^2\epsilon_0^2} \qquad (1.1)$$

where the electron mass and charge are $m = 9.109 \times 10^{-31}$ kg, and $e = 1.602 \times 10^{-19}$ C, respectively; Planck's constant, $h = 6.626 \times 10^{-34}$ J s, and the permittivity of free space $\epsilon_0 = 8.854 \times 10^{-12}$ C V^{-1} m^{-1}.

▷
A useful unit of length on the atomic scale.

(c) Given that E may also be expressed in terms of a_0, the radius of the first Bohr orbit in the hydrogen atom ▷, in the form

$$E = -\frac{e^2}{8\pi\epsilon_0 a_0},$$

use Equation (1.1) to show that $a_0 = \epsilon_0 h^2/(\pi m e^2)$.

1.2   Symbols and more rules

So far we have been dealing with manipulations on numbers using the four basic operations of addition, subtraction, division and multiplication. In order to prepare the ground for a further exploration in the use of symbols to represent numbers, several of the earlier examples and problems have been couched in terms of numbers represented by x, y, a or b, where any of these symbols can be interpreted in terms of an integer or real number, depending upon the context. The obvious advantage in this move away from the need to specify explicit numbers in an expression is that it enables us to make mathematical statements which are quite general: a feature which is developed further in the next chapter on functions.

> See Table 2.3.

A good illustration of this kind of generalisation is seen in the discussion of heat capacities by Alberty and Silbey ▷, where the variation in molar heat capacity at constant pressure in the case of gaseous CO_2 can be fitted by the formula

$$C_p/\text{J K}^{-1}\,\text{mol}^{-1} = 26.648 + 42.262 \times 10^{-3}\,(T/\text{K})$$
$$- 142.4 \times 10^{-7}\,(T^2/\text{K}^2)$$

for $T/\text{K} = 300$ to 1500. However, using the notation of algebra, a formula can be given in which, instead of the explicit real number coefficients, we can use symbols – say a, b and c. Thus, the formula becomes $C_p/\text{J K}^{-1}\,\text{mol}^{-1} = a + bT + cT^2$, which is applicable to any gaseous species. This move from the specific to the general is provided by the tools of algebra, and yields a most useful and compact expression. For the rest of this chapter we shall be concerned with the rules for manipulating algebraic expressions.

> Degree means the highest power or index of the variable.

The specific heat example above gave us a polynomial expression of degree 2 ▷ in the variable T (in the use of such expressions, a, b, and c are thought of as constants, as the main focus of interest is in the variation of C_p with temperature and not a, b and c with species). Polynomial expressions of degree one, two and three, ... are usually termed *linear*, *quadratic*, *cubic*, ... expressions, respectively; in this terminology, a constant is just a polynomial of degree zero.

Handling algebraic expressions requires a little care, especially where signs are concerned, and it is helpful to give the rules for combining polynomial expressions (this is sufficient for most purposes because, when we meet non-polynomial expressions, they can be manipulated in a similar way).

Suppose the expressions $1 + x$ and $x^2 + 1$ are multiplied together. In order to preserve the integrity of the two expressions, brackets are introduced prior to collecting up terms in the product: $(1 + x)(x^2 + 1)$. Now if the first and second expressions are thought of as the objects $\square$ and $\triangle$, respectively, the expansion of the brackets can be accomplished in a simple set of steps using the *distributive* rule ▷:

> Either in the form
> $a(b + c) = ab + ac$ or
> $(a + b)c = ac + bc$.

$$(1 + x)(x^2 + 1) \equiv \square\triangle = \square(x^2 + 1) = \square \cdot x^2 + \square \cdot 1$$
$$= (1 + x) \cdot x^2 + (1 + x) \cdot 1$$
$$= x^2 + x^3 + 1 + x,$$

where, in the last step, the brackets are not used because there is no ambiguity when adding polynomial expressions.

Great care must always be taken with signs. For example,

$$\square - \triangle = (1+x) - (x^2 + 1)$$
$$\equiv (1+x) - 1 \cdot (x^2 + 1) = 1 + x - x^2 - 1,$$

where the $-\triangle$ term is expanded using the first form of the distributive rule with $a = -1$. Similarly, when expressions involving multiplication, division and addition are being manipulated, illogical errors are avoided by the proper use of brackets. For example, $2 + 6$ may be written as $2(1 + 3)$, or $(4 + 2x)/6$ may be simplified to $2(2 + x)/(3 \cdot 2) = (2 + x)/3$, by cancelling the common factor 2 in numerator and denominator. Furthermore, any expression common to both denominator and numerator can be cancelled, so long as it does not take the value zero.

In some circumstances, especially those involving factors containing fractional exponents, it may be helpful to multiply numerator *and* denominator by a suitably chosen common factor, as is seen in the following example.

Worked example

1.6 Simplify

$$\frac{2}{(\sqrt{3} - \sqrt{2})}$$

Solution It may be converted to an expression that has no square root factors in the denominator by multiplying numerator and denominator by $(\sqrt{3} + \sqrt{2})$:

$$\frac{2}{(\sqrt{3} - \sqrt{2})} \cdot \frac{(\sqrt{3} + \sqrt{2})}{(\sqrt{3} + \sqrt{2})} = \frac{2(\sqrt{3} + \sqrt{2})}{3 - 2} = 2(\sqrt{3} + \sqrt{2}).$$

>
Levine p. 552.

Problem 1.7

The manipulations required in working out both parts of this question are required in calculating the π-orbital energies of butadiene ▷.

(a) Convert $\dfrac{1}{(\sqrt{5} + 1)}$ to an expression that does not have square root factors in the denominator.

(b) Show that $(1 - \sqrt{5})^2 = 6 - 2\sqrt{5}$, and thus simplify $(3 - \sqrt{5})^{\frac{1}{2}}/\sqrt{2}$.

(c) Simplify $\dfrac{(1 - x)}{(1 - \sqrt{x})}$.

The simplification of more involved algebraic expressions requires repeated applications of the distributive rule for introducing or expanding a bracket, usually with the initial formation of the lowest common denominator (where appropriate).

Worked example

1.7   Simplify:

$$\frac{(1+x)^2(x^2+1)}{(1-x)(1+x)} - \frac{1}{1-x}.$$

Solution

▷

It is necessary that $x \neq -1$, as it is a meaningless operation to cancel zero factors.

To appreciate what is involved, think of each expression in brackets as an object: for example, $\square = 1+x$, $\triangle = x^2+1$, and $\diamond = 1-x$. Thus, by cancelling the common factor $(1+x)$ in the first rational expression ▷ and then forming a common denominator, we obtain

$$\frac{\square^2\triangle}{\diamond\square} - \frac{1}{\diamond} = \frac{\square\triangle}{\diamond} - \frac{1}{\diamond} = \frac{\square\triangle - 1}{\diamond} = \frac{(1+x)(x^2+1) - 1}{\diamond}$$

$$= \frac{(x^3+x^2+x)}{(1-x)} = \frac{x(x^2+x+1)}{(1-x)}.$$

> **Problem 1.8**
>
> Simplify:
>
> (a) $x - \dfrac{x}{(x+1)}$,   (b) $2 + \dfrac{1}{(x-1)} - \dfrac{1}{(x+1)}$.

The method for dealing with multiplying out two bracketed quantities is described above. In an extension of this procedure, it is a useful exercise to examine the terms that are produced in the expansion of $(1+x)^n$, where n is a positive integer (the techniques for dealing with negative integer and rational n are described later). Since each bracket contains two terms, and there are n brackets multiplied together, there are 2^n terms in all: for example, with $n = 2$ we have

$$(1+x)^2 = (1+x)(1+x) = 1\cdot 1 + 1\cdot x + x\cdot 1 + x\cdot x \qquad (2^2 \text{ terms})$$
$$= 11 + 1x + x1 + xx \tag{1.2}$$
$$= 1 + 2x + x^2, \tag{1.3}$$

where the · has been removed to display symbol strings: for example 11 does not mean eleven, but two ones in juxtaposition (multiplied together). Similarly, for $n = 3$,

$$(1+x)^3 = (1+x)(1+x)^2 = (1+x)(1 \cdot 1 + 1 \cdot x + x \cdot 1 + x \cdot x)$$
$$= 1 \cdot 1 \cdot 1 + 1 \cdot 1 \cdot x + 1 \cdot x \cdot 1 + 1 \cdot x \cdot x + x \cdot 1 \cdot 1 + x \cdot 1 \cdot x$$
$$+ x \cdot x \cdot 1 + x \cdot x \cdot x$$
$$= 111 + 11x + 1x1 + 1xx + x11 + x1x + xx1 + xxx \qquad (1.4)$$
$$= 1 + 3x + 3x^2 + x^3$$
$$= (1 + 2x + x^2) + x(1 + 2x + x^2). \qquad (1.5)$$

A comparison of Equations $(1.2), (1.4)$ shows how the coefficients of powers of x in the cubic expression are derived from those in the quadratic:

x^0	x^1	x^2	x^3
1	2	1	
	1	2	1
1	3	3	1

It is clear that the coefficients of the powers of x for the cubic can be obtained directly from the pattern of coefficients for the quadratic by displacing the latter set one power of x to the right and then adding the two sets of coefficients.

The significance of these patterns of numbers in the expansion of $(1 + x)^n$ becomes apparent in several different areas of chemistry, notably in the study of nuclear magnetic resonance (n.m.r.) spectroscopy. For example, three equivalent protons, each having one of two spin states, give rise to 2^3 spin arrangements, the enumeration of which is described in the above analysis for the expansion of $(1 + x)^3$, where the symbols 1 and x in Equation (1.4) represent the two spin states of the proton. As noted in Atkins $\triangleright$, the intensities of the lines in the n.m.r. spectrum are given by successive coefficients of x^k in the expansion of $(1 + x)^3$.

$\triangleright$
Atkins, p. 634.

Problem 1.9

From the coefficients of $x^k, k = 0, 1, 2, 3$, for $(1 + x)^3$ given above, write down the expansion of $(1 + x)^4$.

$\triangleright$
Usually termed *binomial* coefficients.

It is quite easy to generate the formula for the coefficients $\triangleright$ of x^k in the expansion of $(1 + x)^n$, since each of the 2^n terms corresponds to an n-fold product of 1's and x's containing either 0, 1, 2, ... or n occurrences of x. Starting with a string of n 1's, there are n positions where a 1 can be replaced by an x, i.e., the coefficient of x in the expansion of $(1 + x)^n$ is n. Two replacements of 1 in the string can be made in $n(n - 1)/2$ ways – n possibilities for the first replacement, thus leaving $(n - 1)$ possibilities for the second; however, the same arrangement is achieved in two ways, and hence there is a double counting because if a and b designate the first and second choices for x then $\cdots 1a11b1 \cdots$ and $\cdots 1b11a1 \cdots$ correspond to the same arrangement $\cdots 1x11x1 \cdots$. Continuing with three 1's replaced by x's, there are $n(n-1)(n-2)/(3 \cdot 2 \cdot 1)$ different arrangements because, now, the first x can be chosen in three ways (since three x's appear in the string); the second in two ways, and the last x in one way. Extending these ideas to k replacements of 1 by x leads to $n(n-1) \cdots (n-k+1)/(k \cdot (k-1) \cdots 1)$ different distinguishable arrangements. Thus, in the expansion of $(1+x)^n$, the coefficient of x^k (the binomial coefficient ^nC_k) is $n(n-1) \cdots (n-k+1)/k \cdot (k-1) \cdots 1$, i.e.,

$$
^nC_k = \frac{n(n-1) \cdots (n-k+1)}{k \cdot (k-1) \cdots 1} \cdot \frac{(n-k)(n-k-1) \cdots 1}{(n-k)(n-k-1) \cdots 1}
$$

$$
= \frac{n!}{k!(n-k)!} \equiv \binom{n}{k} = \frac{n(n-1) \cdots (n-k+1)}{1 \cdot 2 \cdot 3 \cdots k}
$$

which is written using the *factorial* notation ▷.

$▷$
$n! = n(n-1)(n-2) \cdots 1$
is just the product of the n integers from n to 1, inclusive, and reads n factorial. $0! = 1$ by convention.

$▷$
Smart and Moore, Chapter 3.

This kind of analysis is also relevant to several different kinds of chemical situations where, in particular, the calculation of entropy effects needs to be carried out. In solid state chemistry, for example, a sample of pure crystalline NaCl is described in terms of two interpenetrating face-centred cubic arrays of Na^+ and Cl^- ions (in the ionic model of NaCl). Real crystals may possess defects of varying kinds: a Schottky defect ▷ arises when there is an absence of one cation and one anion. If there are k Schottky defects, then the cation vacancies will arise in many ways: in fact the same number of ways as there are anion vacancies, in order to conserve charge neutrality. (A cation vacancy is a site in the crystal where there is an absence of a cation.)

Problem 1.10

Suppose there are N cation sites in a pure crystal of NaCl. Let this structure be represented by a string of N 1's:

$\cdots 1111111111 \cdots$.

Suppose there are k Schottky defects, i.e., absences of k Na^+ species. Let the absence of a cation be represented by the replacement of 1 in the string by x.

(a) How many 1's are replaced by x's?

(b) Using the result above, give the number of ways in which k cation defects can be distributed in the crystal.

(c) Give the number of ways of distributing the k anion vacancies.

(d) Give the total number of ways in which k Schottky defects can be distributed.

1.3   Simple polynomial equations

This chapter has so far concentrated on numbers and symbols and their manipulation. A simple polynomial equation is of the form $P(x) = 0$, where $P(x)$ is a polynomial of degree q: for example, with $q = 2$,

$$x^2 + 5x + 6 = (x + 2)(x + 3) = 0,$$

where the second equality shows the polynomial in factored form. A polynomial equation has a solution if values of the unknown x can be found such that, on substituting into the polynomial expression, the resultant value of zero is obtained. In general terms, it is possible to find up to q values of x satisfying the equation. In the present example, -2 and -3 are the only values of x for which the polynomial has the value zero. These values form the solutions (or roots) of the polynomial equation. However, for

$$x^2 + 2x + 1 = (x + 1)(x + 1) = 0$$

only $x = -1$ is a realizable root, while for

$$x^2 - 2x + 2 = 0$$

> For the three examples, $S_1 = \{2, 3\}$, $S_2 = \{1\}$, $S_3 = \emptyset$, where $\emptyset$ is the null or empty set.

> This is the method of *deflation*.

no real number values of x satisfy the equation. We say that, for these three examples, the *solution* set, S, contains two, one and zero real elements, respectively. A qth degree polynomial equation will have a maximum of q distinct values of x in its solution set $\triangleright$. There are various algebraic ways of finding the elements of the solution set. In some cases it may be possible to factorize the polynomial by finding one or more roots by inspection. Thus, for example, $x = 1$ is a root of the equation $x^3 - 4x + 3 = 0$, thereby enabling the equation to be rewritten in the form $(x - 1)(x^2 + x - 3) = 0$ $\triangleright$; the other two roots may then be found by solving the second-degree polynomial equation $x^2 + x - 3 = 0$. If the inspection method fails to factorize the original polynomial equation then there are always algebraic methods if $q \leq 4$; however, for $q > 4$, it is only possible to determine the elements of the solution set numerically by direct substitution into the equation, systematically incrementing x by a small amount, to determine the regions of x where $P(x)$ changes sign. This process is then iterated, with decreasing size of increment, until a root is located to the required number of decimal places. The corresponding factor then can be removed from $P(x)$ by deflation, to produce a polynomial function of reduced degree as a starting point for locating the next root, and so on. In practice this 'brute force' method may cause acute problems

> See, for example, Press *et al.*, Section 9.5.

when using a numerical procedure programmed on a computer, $\triangleright$ if $P(x)$ contains multiple, or closely spaced, roots. The Newton–Raphson method, described in Chapter 5, will also be seen to provide a way of locating one or more roots of a polynomial equation; as with all numerical procedures, however, convergence to a given root is not always guaranteed.

In the case of a quadratic or cubic polynomial equation, formulae are available which yield the values of x lying in the solution set. The former case is easy to treat, but there are two approaches for dealing with the cubic

> Beyer, pp. 9–12.

equation, both of which are somewhat involved $\triangleright$.

▷

After dividing each
coefficient in the
equation by the
coefficient of x^2, if
necessary.

The solution of the quadratic $x^2 + bx + c = 0$ can be achieved by the following procedure ▷:

$$x^2 + bx + c = (x + b/2)^2 - (b^2/4 - c) = 0.$$

On letting $(x + b/2)$ and $(b^2/4 - c)^{\frac{1}{2}}$ be represented by X and A, respectively, the original polynomial equation becomes

$$X^2 - A^2 = 0$$

i.e., the left-hand side vanishes if $X = A$ or $X = -A$, which implies that

$$(x + b/2) = \pm(b^2/4 - c)^{\frac{1}{2}} = \pm\frac{1}{2}(b^2 - 4c)^{\frac{1}{2}}$$

$$\Rightarrow x = -\frac{b}{2} \pm \frac{1}{2}(b^2 - 4c)^{\frac{1}{2}}. \tag{1.6}$$

It is now possible to see why the solution set for a quadratic equation may contain between zero and two elements: if $b^2 > 4c$ or $b^2 = 4c$ then there are either two values or one value of x in the solution set, respectively; however, if $b^2 < 4c$ then S is the null set because there is no real number that corresponds to the square root of a negative number – thus there is no real x satisfying the quadratic equation.

Problem 1.11

Find the solution set for each of the following equations:

$2x - 1 = 0, x^2 - 3x + 1 = 0, (x + 1)(x^2 - 5x + 4) = 0,$
$x^2 + 2x + 10 = 0.$

▷

See Atkins, p. 432.

▷

See Problem 1.6(c) for
the definition of a_0.

Problem 1.12

The $3s$ atomic orbital of hydrogen has two radial nodes, each giving rise to a spherical nodal surface ▷. The equation giving the location of these nodes is $\rho^2 - 6\rho + 6 = 0$, where ρ is proportional to the distance, r, from the nucleus. Find the solution set for this quadratic equation and, using the relation $r = 3\rho a_0/2$, determine the two values of r which locate the nodes ▷.

▷

See Atkins, p. 498 where
w is identified with
$(\alpha - \epsilon)/\beta$.

Problem 1.13

The energies of the π-molecular orbitals in butadiene may be expressed as $\epsilon = \alpha - \beta w$, where w is in the solution set of $w^4 - 3w^2 + 1 = 0$, and α, β are parameters in the Hückel model ▷.

(a) Transform the quartic equation in w by using the substitution $w^2 = x$, and show that $x = (3 \pm \sqrt{5})/2$.
(b) Use the result from Problem 1.7 but to determine the four values for w (it is not necessary to approximate the value of $\sqrt{5}$).
(c) Given that α and β both have negative values, determine expressions for the two lowest energy π-molecular orbital energies.

1.4   A first look at complex numbers

As seen above, the solution set for the quadratic equation $x^2 - 2x + 2 = 0$ is empty because the formula for the quadratic yields x values of $1 \pm (-4)^{\frac{1}{2}}/2$, and neither value of x is real. Solutions can be found to this kind of equation by extending the number system to include *complex* numbers, which are different in nature from ordinary real numbers. A more detailed coverage of complex numbers is given in Chapter 8 but, in the meantime, it is sufficient to say that if i is defined as $\sqrt{-1}$, then the solution set for $x^2 - 2x + 2 = 0$ contains $1 \pm i$ ▷. Numbers of the form $x = a + ib$ are termed *complex* numbers; if $b = 0$ then x is a real number, and if $a = 0$ then x is an *imaginary* number. The imaginary number i may be treated in algebraic expressions like any real number, with the additional property that $i^2 = -1$. Also, if $x = a + ib$ then the *complex conjugate* of x is designated by $x^* = a - ib$.

▷
From law (5) of the properties of exponents, $(ab)^{\frac{1}{2}} = a^{\frac{1}{2}}b^{\frac{1}{2}}$, and so taking $a = -1$ and $b = 4$ yields the stated result.

Problem 1.4

(a) Find the solution set, S, of $x^2 + 2x + 10 = 0$. For each element of S, write down xx^*.
(b) Express $x = (1 + 2i)(3 - i)$ in the form $x = a + ib$, remembering that $i^2 = -1$; hence determine x^2 and x^3.

Summary: This concludes the introductory chapter on numbers, mathematical rules and symbolism, together with a brief excursion into algebra, and an even briefer excursion into complex numbers (a more detailed discussion of the latter follows in Chapter 8).

The next chapter builds on the above discussion of polynomials to explore the idea of function – a most important mathematical entity that pervades many areas of chemistry in one form or another (especially thermodynamics).

2 Functions of a single variable

Objectives

This chapter

- describes how the mathematical idea of function can be used to represent the value of a target chemical property in terms of a rule or prescription

- demonstrates the importance of defining the set of values of the independent variable for which the rule or prescription is valid

- introduces the nature and importance of the modulus, exponential, hyperbolic and logarithm functions, together with selected trigonometric functions

- provides examples and problems which relate the ideas in the chapter to specific chemical situations

In chemistry we frequently investigate how a particular property of a system changes when either the concentration of a species, the volume, the pressure or the temperature changes: for example, the pH of an acid or base in water varies according to the activity of H^+ species in the solution; the value of an enthalpy change associated with a chemical reaction or physical change (phase transition) varies with temperature; likewise, the equilibrium constant for a chemical reaction varies with temperature, as does the rate constant for a reaction. There are many situations when the property of interest depends upon two or more variables. For example, the vibrational motion of the nuclei in the water molecule may be described in terms of the variation of three coordinates (two bond lengths and the interbond angle); the electron in a hydrogen atom orbital requires three coordinates to describe its position in three-dimensional space and, finally, the rate of reaction between H_2 and Br_2 depends upon the concentrations of these species and that of HBr, i.e., three variables.

2.1   The idea of function

In all the above examples, the value of the target property of interest can be related to values of properties on which it depends by means of a simple rule or prescription. Such rules or prescriptions are intimately related to the mathematical ideas of function which provide a formalism for describing interrelationships between a set of variables (the input, or independent variables) and the value of the target property (the output, or dependent variable). Since we are concerned with the link between *values* of target properties and the *values* of the independent variable(s), the input and output data are just numbers (for the present, at least). The context of the equation relating x to y must, of course, be such that the relation is meaningful, and so the physical properties associated with the two variables must be such that the units are consistent. In practice this means that coefficients in the formula for y carry appropriate units. For example, in the formula $C_p = a + bT + cT^2$, a, b and c carry units of J K^{-1} mol^{-1}, J K^{-2} mol^{-1} and J K^{-3} mol^{-1}, respectively ▷. The formula itself does not depend on any particular choice of units; thus, when we write $y = f(x)$, we mean that given a value for the real (independent) variable x, in the appropriate units, we can produce *one* value for the physical property $f(x)$ – here called y (the dependent variable) – using a simple prescription, rule or recipe:

▷
Alberty and Silbey, p. 52

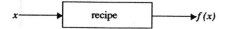

The rule, given the name f, may be described by a simple formula; but, in the example cited above, concerned with the variation of the molar heat capacity, C_p, with temperature, it is common practice to use observed experimental data to deduce a best guess for the coefficients a, b and c in the formula linking the two properties. The procedures for justifying the formula itself and for fitting the experimental data to the formula are discussed in later chapters.

Simple algebraic functions

The form of a function can be described in terms of

- the formula $y = f(x)$, involving a rule or prescription for obtaining one $f(x)$ value from each value of x

- the graph of $f(x)$ (as ordinate) plotted against x (as abscissa)

▷
The symbol: means 'such that'.

- the set of ordered pairs $\{(x, y) : y = f(x)\}$ ▷

- a mapping from the set, $\mathbb{R}$, of real numbers x to another set, $\mathbb{R}'$, of real numbers $f(x)$.

In all these descriptions, the set of x values for which $f(x)$ is defined forms the *domain*; the set of associated $f(x)$ values forms the *codomain* or range. Although the first and second descriptions are more intuitive, it is important to become more familiar with all four modes of description. The fourth description, in particular, is especially useful in describing the geometrical transformations which are important in defining the symmetry properties exhibited by a molecule (see Chapter 13).

To illustrate the different ways of understanding the notion of function, consider $y = f(x)$ defined by the rule $f(x) = x + 2$. This rule is clearly valid for all real numbers x, i.e., the domain is the set $\mathbb{R}$. Furthermore, since all real numbers, y, are obtained after applying the rule, the codomain is also the same set, $\mathbb{R}$, as the domain. The graph of the function is given in Figure 2.1. The mapping describing the function is shown schematically in Figure 2.2.

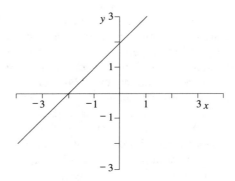

Figure 2.1
The (partial) plot of the function $y = x + 2$

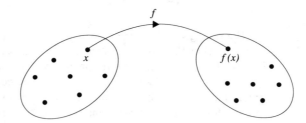

Figure 2.2
The schematic representation of a function defined in terms of the mapping from $\mathbb{R}$ to $\mathbb{R}$

Worked example

2.1 For the function shown in Figure 2.1, give the values of x for which

(a) $f(x) < 0$,   (b) $f(x) \geq 0$,   (c) $0 \leq f(x) < 2$.

Solution   (a) $f(x)$ is negative and non-zero (that part of the graph below the x-axis) for all values of x such that $x < -2$, i.e., for all x in the set $\{x : x < -2\}$.
(b) $f(x)$ is positive or zero (that part of the graph lying on and above the x-axis), for $x \geq -2$, i.e., for all x in the set $\{x : x \geq -2\}$.
(c) Here the set of x values such that $f(x)$ is greater than or equal to zero, but less than 2, lie in the set $\{x : -2 \leq x < 0\}$.

In all parts of Example 2.1, we are looking for a particular *subset* of $\mathbb{R}$ in which values of $f(x)$ are constrained to satisfy some requirement. This idea will be extended later in this chapter to provide a way of defining $f(x)$ in terms of different formulae for different subsets of $\mathbb{R}$.

Although, as noted above, it is intuitively suggestive to display a function in graphical form, it is better to work with the basic formula because, in extending the concept of function to two or more independent variables, it becomes increasingly difficult to present meaningful geometrical representations of such objects. A classic example is an atomic orbital function, depending upon three position coordinates, which can only be visualized in four dimensions (the value of the function is the dependent variable).

The inverse function

The convention adopted here, which is widely followed, is to refer simply to $y = f(x)$ (or even $f(x)$ itself) as the function when, of course, this is strictly not the case. In addition, we shall refer to $y = f(x)$ as an equation for determining y from x and, in this situation, y is said to be the subject of the equation. It may happen, however, that we need to express x in terms of y by changing the subject of the equation. This is often straightforward, and involves the use of the rules of algebra to make the necessary rearrangement. The resulting expression may or may not be a function as we have defined it here, and the following examples will make this situation clear. However, if the new expression defines a function, g, then it provides a way of defining an *inverse* function which reverses the effect of f on the variable x: a property most easily seen by using the codomain of f as the domain for the new function.

Worked example

2.2 (a) For the function $y = f(x) = x + 2$, defined for all x (see Example 2.1), make x the subject of the equation. Confirm that the new formula defines a function which can be identified as the inverse function of f.

Solution It follows that $x = y - 2$ when x is made the subject of the equation. This new expression gives us another function because, for any value of the independent variable, now named y, the value of x is uniquely determined. Let this new function be given the name g, so that $y = g(x) = x - 2$ on renaming the dependent and independent variables as y and x, respectively $\triangleright$.

▷
The name for a variable is not special.

▷
Image means the value of $f(a)$, or the value into which a is mapped.

A given number a, when substituted in the formula for $f(x)$, yields the number $y = a + 2$ (the image $\triangleright$ of a under f). The number $a + 2$, when substituted for x in the formula defining $g(x)$, then yields $a + 2 - 2$, i.e., a. It is clear, therefore, that g reverses the effect produced by f: this is why g is termed the *inverse* function of f, and often given the name f^{-1}.

(b) Given the function $y = x^2$, which is valid for all x, make x the subject of the equation and check whether an inverse function can be defined.

Solution On rearranging the given equation it follows that $x^2 = y$, from which we deduce $x = \pm\sqrt{y}$; or, in conventional notation, with y named as the dependent variable, $y = \pm\sqrt{x}$. This latter expression does not define a function because, for a given value of x, there are two associated values of the dependent variable, y. It is usual in this situation to consider the two expressions $y = \sqrt{x}$ and $y = -\sqrt{x}$ separately which, taken with a domain of positive or zero x, define the inverse functions of $y = x^2$ (for zero and positive x) and $y = x^2$ (for negative x), respectively. This example illustrates the important point that a function always requires a formula *and* a domain in order to be defined properly.

Worked example

2.3 The volume, V, of a sphere and its radius, r, are related by the rule

$$V = g(r) = \frac{4}{3}\pi r^3$$

where g is the name of the rule that describes the manipulation of the independent variable (r) to produce the value of the dependent variable (V). Find the inverse function of $g(r)$.

Solution Here, $g(r)$ is a function with domain equal to the set of real positive numbers $\triangleright$, and an inverse function can be defined in the form $r = (3V/4\pi)^{\frac{1}{3}}$ because there is only one real cube root.

$\triangleright$
$r = 0$ does not define a sphere!

More on the domain

In the examples given so far, the importance of specifying a function by means of both a rule and the values of x for which the rule is applicable (the domain) have been stressed. In some situations the domain of a function is not at all obvious, and the examples and problems which follow illustrate some of the techniques that have to be used.

Worked example

2.4 Give the domain of $f(x) = (x^2 - 4)^{\frac{1}{2}}$ and evaluate $f(4)$.

Solution For $f(x)$ to be defined, $(x^2 - 4)$ must be greater than or equal to zero, that is $(x^2 - 4) \geq 0$. By adding 4 to each side of this inequality, we see that $x^2 \geq 4 \Rightarrow x \geq 2$ or $x \leq -2$:

$$\xleftarrow{\quad}\underset{-2}{\vdash}\quad\underset{0}{}\quad\underset{+2}{\vdash}\xrightarrow{\quad} x$$

The domain of $f(x)$ is thus given by all those values of x for which either $x \geq 2$ or $x \leq -2$, as shown above by the thick portions of the number line: that is, the permitted values of x lie in the subset of numbers from $+2$ to an indefinitely large positive number (usually termed infinity, and designated by ∞), or in the subset of numbers from an indefinitely large negative number $(-\infty)$ to -2. These subsets are usually described using the notation $x \in [2, \infty)$ or $x \in (-\infty, -2]$, where [or] means that the appropriate end-point *is* included in the domain; open brackets (or) mean that this end-point is excluded ▷. If the two subsets are designated by $S_1 = \{x : x \in [2, \infty)\}$ and $S_2 = \{x : x \in (-\infty, -2]\}$, then the domain of $f(x)$ is given by the totality of the values of x in both S_1 and S_2: usually written as $\mathbb{S} = S_1 \cup S_2$, where $\cup$ is the *union* ▷ operation on the sets. $\mathbb{S}$ is, of course, the set of numbers left when the subset $\{x : x \in (-2, 2)\}$ is removed from $\mathbb{R}$.

$x = 4$ is in the domain of $f(x)$, and thus $f(4) = 12^{\frac{1}{2}} = 3.464$.

Finally, the behaviour of $f(x)$, as x becomes indefinitely large (in either direction) is called its *asymptotic behaviour*. Thus, the asymptotic values of $f(x) = (x^2 - 4)^{1/2}$ are infinite.

▷

Any very large positive or negative number is acceptable here.

▷

A kind of summation – more akin to inclusion.

Problem 2.1

(a) Give the domain of the functions defined by $f(x) = x + (x^2 - 4)^{\frac{1}{2}}$, and $g(x) = 1 + 1/x$.

(b) Evaluate $f(4)$, and give the asymptotic values of $g(x)$ as x becomes indefinitely large in both positive and negative senses.

Worked example

2.5   A function is defined by

$$f(x) = \left\{ \frac{x}{(1 - x)} \right\}^{\frac{1}{2}}.$$

Give the values of x for which the rule defining $f(x)$ is meaningful, and write down the values of $f(0), f(\frac{1}{2})$.

Solution   For the formula to be meaningful, the value of $x/(1 - x)$ must be greater than or equal to zero, with $x \neq 1$:

$$\frac{x}{1 - x} \geq 0, \quad (x \neq 1).$$

Now if both sides of this inequality are multiplied by the positive number $(1 - x)^2$ then, as seen in Chapter 1, the sense of the inequality is unchanged. Hence, $x(1 - x) \geq 0$, and the left-hand side is positive or zero if either x and $(1 - x)$ are both positive or both negative: that is, $x \geq 0$ and $(1 - x) > 0$ or $x \leq 0$ and $(1 - x) < 0$ ▷. The values of x satisfying either of these inequalities are best found by looking at the diagram below, where the signs of the factors x and $(1 - x)$ for various values of x are indicated.

▷

The situation where $x = 1$ is excluded by the choice of inequalities.

$$x \geq 0$$
$$(1-x) > 0$$
$$x \leq 0$$
$$(1-x) < 0$$

$$-3 \quad -2 \quad -1 \quad 0 \quad 1 \quad 2 \quad 3$$

The first solution set, S_1, defining part of the domain, contains those x for which $x \geq 0$ and $x < 1$; the second solution set, S_2, which requires $x \leq 0$ and $x > 1$, is empty. Thus, the domain is given by the union of S_1 and S_2, which is just $S_1 = \{x : x \in [0, 1)\}$, and $f(0) = 0, f(\frac{1}{2}) = 1$.

Problem 2.2

For the function f defined by

$$f(x) = \{(2 + x)/(1 - x)\}^{\frac{1}{2}},$$

give the domain for $f(x)$, and write down the values of $f(0), f(\frac{1}{2})$ and $f(-1)$.

Problem 2.3

If

(a) $f(x) = x^2 - 3x$, find $f(-1), f(0), f(\frac{1}{3})$;

(b) $p(x) = x/(1 + x)$, write down $p(1/x), p(2 - t)$;

(c) $g(x) = 1/x$, show that $\dfrac{g(x + h) - g(x)}{h} = -\dfrac{1}{x(x + h)} (h \neq 0)$;

(d) $T(x) = \sqrt{1 - x^2}$, find $T(\frac{1}{4}), T(-\frac{1}{4}), T(0)$.

For the functions defined in (a)–(d) above, give the domain and codomain, and find the asymptotic values of $f(x), p(x)$, and $g(x)$ as x becomes indefinitely large in a positive sense.

The examples and problems discussed so far relate to functions described by a simple formula or rule for which the independent variable is a real number. In chemistry there are many instances where a particular property is known or defined only for a subset of the integers, $\mathbb{I}$. For example,

- the electronic energy levels for the hydrogen atom are given by the formula $E = -R/n^2$; $n = 1, 2, 3, \ldots$ The domain is given by the set of counting numbers (a subset of $\mathbb{I}$).
- any observed or postulated atomic property, such as ionization energy, atomic radius or electronegativity is a function defined with domain equal to a subset of the positive integers $\{1, 2, \ldots, 109\}$, corresponding to the atomic numbers of the (known) elements.

In the former example, the discrete nature of the energy level pattern is usually emphasized; but in the latter examples, the point values of atomic properties

▷

Greenwood and
Earnshaw, Chapter 2.

are usually joined up to demonstrate variations and trends ▷. Plots of these properties must not be viewed as displaying a functional form, because there is no mathematical justification for joining the points (interpolation), since we would then be implying that the independent variable takes values outside 𝕀.

Problem 2.4

▷

The centre of mass is
the 'balancing point' at
distances r_1 from A and
r_2 from B, such that
$r_1 m_A = r_2 m_B$ and
$r_1 + r_2 = R$, the
internuclear separation
(see Alberty and Silbey,
Section 10.13 and
Example 10.13).

The energy levels for the rotational motion of a diatomic molecule AB about an axis passing through the centre of mass ▷ are given by

$$E = J(J+1)\frac{h^2}{8\pi^2 I}$$

where $I = m_A r_1^2 + m_B r_2^2$ is the moment of inertia, and J is the rotational quantum number taking values $0, 1, 2, \ldots$

(a) Solve the two equations given in the marginal note to show that $r_1 = m_B R/(m_A + m_B)$ and $r_2 = m_A R/(m_A + m_B)$.

(b) Substitute for r_1 and r_2 in the equation for I above, and derive the alternative expression $I = \mu R^2$, where $\mu = m_A m_B/(m_A + m_B)$.

(c) Given that $I = 2.644 \times 10^{-47}$ kg m^2 for ^{1}H^{35}Cl, calculate the values of $E(J)$ for $J = 0, 1, 2, 3, 4$.

(d) What is the domain of the function $E(J)$?

(e) Give a graphical display of the function $E(J)$, using 5 cm $\equiv$ 4.0 $\times 10^{-21}$ J.

Functions as prescriptions

The ideas of function discussed so far are restricted in the sense that one rule or prescription is associated with the domain. We need to extend these principles in order to encompass more general kinds of function which are described by a different formula or rule for different parts of the domain. Functions of this kind are fairly common in chemistry: for example, if the variation of either enthalpy or heat capacity is displayed as a function of temperature, sudden increments in value are observed at critical temperatures corresponding to a change in phase ▷. Other physical properties also show sharp changes in general trends, but without characteristic breaks, at certain critical temperatures. Typical examples include the sudden changes in

▷

See p. 96, Alberty and
Silbey, for the variation
in heat capacity of SO$_2$
with temperature.

▷

See Barrow, Figure 18.7.

▷

See Shriver et al., Figure
B18.3, p. 764.

- rate constant as a function of temperature for a (chain) reaction that reaches an explosive stage where change of phase ▷ is not a determining factor;

- magnetic susceptibility of a material containing unpaired electrons (NiO) at critical temperatures associated with the onset of antiferromagnetic behaviour ▷;

See Shriver *et al.*,
Figure 18.17, p. 762.

- electrical conductivity of a material (V_2O_3) at a temperature corresponding to a change from metallic to semiconducting behaviour ▷.

In situations where the relation between physical properties suffers sharp changes or distinct breaks, it is necessary to describe the function relating the independent and dependent variables in terms of a prescription with different rules for different subsets of values for the independent variable.

Problem 2.5

Consider the *modulus* function $y = |x|$, defined by the prescription:

$$|x| = \begin{cases} x, & \text{if } x \geq 0 \\ -x, & \text{if } x < 0. \end{cases}$$

(a) Sketch the form of this function for x values of $\pm 3, \pm 2, \pm 1, 0$, and indicate the values of x for which $|x| \leq 1$.

(b) Using values of x equal to 0, 1, 2, 3, 4, 5, sketch the form of the function $y = f(x)$, where

$$f(x) = \begin{cases} (x/3) + 1, & \text{if } 0 \leq x < 3 \\ (2x/3), & \text{if } 3 \leq x \leq 6. \end{cases}$$

Comment: This graph is similar in nature to the plot of the variation of $\Delta G^{\ominus}$ with temperature for ZnO, as seen in the Ellingham diagram ▷.

(c) Sketch the function

$$f(x) = \begin{cases} x + 1, & \text{if } x < 2 \\ x, & \text{if } x \geq 2 \end{cases}$$

for the same values of x as given in part (a).

See Shriver, *et al.*,
p. 278.

See Barrow, Section
19.3.

The use of functions in the form of a prescription also arises in the theoretical modelling of chemical processes. For example, in the collision theory of reactive gas-phase reactions ▷, the following two simple prescriptions have been used to define the reactive cross-section area, $\sigma_{AB,R}$, which determines the temperature dependence of the rate constant in the reaction between A and B molecules:

$$\sigma_{AB,R} = \begin{cases} 0, & \text{if } \epsilon < \epsilon_0 \\ \sigma_{AB}, & \text{if } \epsilon > \epsilon_0 \end{cases} \qquad \sigma_{AB,R} = \begin{cases} 0, & \text{if } \epsilon < \epsilon_0 \\ \sigma_{AB}(1 - \epsilon_0/\epsilon), & \text{if } \epsilon > \epsilon_0. \end{cases}$$

Here, the collision diameter, σ_{AB}, is independent of the collision enery, ϵ, and ϵ_0 is a constant.

2.2   Exponential functions

Functions of the form $y = b^x$ (b a constant) are termed *exponential* functions with base b. Functions of this form arise naturally in processes involving growth or decay with time. For example, the interest on a bank account; population growth; radioactive decay; the decrease in concentration of H_2 in its reaction with excess Br_2. Different kinds of exponential function arise, depending upon the characteristic time interval used for examining the decay or growth in the physical property; however, the independent variable does not have to be time and, although in many chemical situations this is the case, in thermodynamics, temperature may be a more appropriate independent variable.

There is an exponential function with a very special and unusual base, denoted by e, that permeates chemistry (and mathematics – especially the calculus). The prescription for defining the function $y = e^x$ is given by the unlikely-looking expression

$$y = \left(1 + \frac{1}{N}\right)^{Nx},$$

▷

The details of how such limiting processes are defined are found in Chapter 3 on limits.

▷

This is the *half-life* of potassium $^{44}_{19}K$.

where it is understood that N has an indefinitely large value ▷. For the purposes of motivating the derivation of this special exponential function, we shall consider the radioactive decay process involving $^{44}_{19}K$.

Consider an amount a of radioactive potassium with mass number 44: at the end of a time interval of 22 minutes ▷, the amount of $^{44}_{19}K$ is half of what it was at the beginning of the interval (the remaining half being converted to $^{44}_{20}Ca$), i.e., $a - a/2 = a(1 - \frac{1}{2}) = d$; thus after 44 minutes (two half-lives), the amount of $^{44}_{19}K$ is $d - d/2 = d(1 - \frac{1}{2}) = a(1 - \frac{1}{2})^2$. After x half-lives, there is an amount $a(1 - \frac{1}{2})^x$ of the potassium mass-44 isotope left, i.e., $a/2^x = a2^{-x}$ from the properties of indices. The amount of the potassium isotope at multiples of the half-life is therefore expressed in terms of the exponential function with base 2. The problem with the argument as presented so far is that the loss of electrons (β^- emission) from the potassium nuclei does not occur in a rush after every 22 minutes: there is a continuous emission in which the amount of the potassium isotope decreases steadily with time, i.e., at *any* time, the amount of $^{44}_{19}K$ can be represented as a fraction of the initial amount. Thus if k is the fraction decaying per second and t is the time in seconds from a chosen reference time (taken as zero) then the amount of $^{44}_{19}K$ decaying in time t is kt. Suppose the time interval t is divided into n subintervals of width t/n: then after one subinterval of time there is an amount $d = a - akt/n = a(1 - kt/n)$ left; after two subintervals $d(1 - kt/n) = a(1 - kt/n)^2$ left, and after time t, $a(1 - kt/n)^n$ left.

Now if x is substituted for $-kt$, then we meet the nth power of the polynomial $(1 + x/n)$ that we have already seen in Chapter 1:

$$(1 + x/n)^n = 1 + x + n(n - 1)(x/n)^2/2!$$
$$+ n(n - 1)(n - 2)(x/n)^3/3! + \cdots,$$

but with x/n replacing x. Hence, on dividing the numerator of the third and subsequent terms on the right-hand side by the power of n in the denominator of each term,

$$(1 + x/n)^n = 1 + x + (1 - 1/n)x^2/2!$$
$$+ (1 - 1/n)(1 - 2/n)x^3/3! + \cdots .$$

▷

As noted earlier, further discussion of the techniques briefly described here is taken up in later chapters on limits and power series.

Thus for large n (very small subinterval width), the expression for $(1 + x/n)^n$ is approximated by ▷

$$(1 + x/n)^n = 1 + x + x^2/2! + x^3/3! + \cdots \qquad (2.1)$$

and depends on the value chosen for x. For $x = 1$, the value of $(1 + 1/n)^n$ can be estimated for large n, since then the series becomes

$$1 + 1 + 1/2! + 1/3! + 1/4! + \cdots \qquad (2.2)$$

which sums to a value greater than 2, as the third and subsequent terms are all positive.

Now $1/3!$ is less than $1/(2 \cdot 2 \cdot 1) = 1/2^2$; $1/4!$ is less than $1/2^3$, etc. Hence,

$$(1 + 1/n)^n < 1 + 1 + 1/2 + 1/2^2 + 1/2^3 + \cdots ,$$

and on writing

$$S = 1 + 1/2 + 1/2^2 + 1/2^3 + \cdots = 1 + 1/2\{1 + 1/2 + 1/2^2 + \cdots\},$$

we see that $S = 1 + S/2 \Rightarrow S = 2$ and it follows that $2 < (1 + 1/n)^n < S + 1$ for large n, i.e., $2 < (1 + 1/n)^n < 3$. If explicit substitution is made for the first seven and eight terms in Equation (2.2) representing $(1 + 1/n)^n$ (large n), we find that $2.718\,06 < (1 + 1/n)^n < 2.718\,25$ when the two numbers are truncated to five decimal places. This process can be repeated for pairs of sequential terms progressively further along the series in order to pinpoint the number that is represented by $(1 + 1/n)^n$ for larger and larger n. However, although each pair of numbers obtained in this manner can be seen to be

▷

Each of the two numbers has a finite number of decimal places.

rational ▷, the number obtained in the limiting situation when n becomes indefinitely large is irrational – a fact that cannot be proved here. This number is usually given the symbol e.

Now returning to Equation (2.1), we can substitute $N = n/x$ and write the left-hand side as

$$(1 + 1/N)^{Nx} = \left((1 + 1/N)^N\right)^x,$$

using the laws of indices. In the limit as N becomes indefinitely large ($N \to \infty$), this expression is just e^x. This recipe defines the *exponential*

<div markdown>

▷

The function is sometimes given the name exp, so that $y = \exp(x) \equiv e^x$.

function ▷ with base e, the values of which are always positive, and tend to zero for increasingly negative x. Hence, resubstituting $x = -kt$, the radioactive decay process leads to an amount of $^{44}_{19}\mathrm{K}$ at a time t of ae^{-kt}, where a is the amount of isotopic K at $t = 0$.

Problem 2.6

Use a calculator to determine the values of $y = e^x$ for x in steps of 1 between -5 and 5, and sketch the graph of the function for $-5 \leq x \leq 5$.

Problem 2.7

The number of molecules, n_i, in a given energy state, E_i, is proportional to $e^{-E_i/kT}$, where k is the Boltzmann constant and T the temperature, i.e., $n_i = Ae^{-E_i/kT}$, where A is a constant. For a system of N molecules with two available energy states, E_1 and E_2 ($E_1 < E_2$), show that the fraction of molecules in the upper state, $n_2/(n_1 + n_2)$, is given by

$$\frac{e^{\Delta/kT}}{1 + e^{\Delta/kT}}$$

where $\Delta = E_2 - E_1$. For $\Delta/k = 5.0$ K, sketch the form of $n_2/(n_1 + n_2)$ as a function of kT/Δ for $T = 0, 5, 10, 15, 20, 30, 40, 60, 100$ K, and estimate the limiting value as T becomes indefinitely large.

2.3   The logarithm function

Given $y = b^x$, the logarithm function is defined as $\log_b y = x$, where b is its base. The two bases with $b = 10$ and $b = e$ are in common use, and the shorthand notations of log and ln are used for designating the function names $\log_{10}$ and $\log_e$, respectively. Logarithms to the base e occur more naturally in a chemical context than other bases; however, base 10 logarithms are encountered in defining a scale of acidity through $\mathrm{pH} = -\log_{10} a_{\mathrm{H_+}}$, or the degree of ionization of a weak acid through $\mathrm{pK} = -\log_{10} \mathrm{K}$.

It should be clear from the schematic figure shown in Figure 2.3(a) that the logarithm function $\log_b$ is the inverse of the exponential function b^x (sometimes called the antilogarithm function), as the number x is first mapped to y by the exponential function, and then mapped back to x by the logarithm function, i.e., $\log_b b^x = x$. Furthermore, reversing the order of the mappings by first taking $x = \log_b y$, and then $b^x = b^{(\log_b y)} = y$ (Figure 2.3(b)) yields the useful identity

$$b^{\log_b x} = x$$

$\triangleright$

Changing the name of the variable does not change the nature of the identity.

for logarithms to the base b, after changing the variable name from y to x $\triangleright$. Thus, if the number $f(x)$ is chosen, rather than x, the identity

$$b^{\log_b(f(x))} = f(x)$$

is obtained.

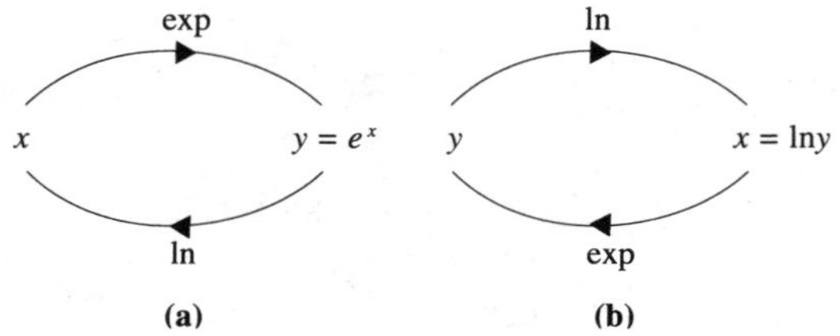

Figure 2.3
The relation between the exponential and natural logarithm functions

The basic properties of logarithms are as follows:

(1) $\log_b(xy) = \log_b x + \log_b y$
(2) $\log_b(x^n) = n\log_b x$
(3) $\log_b(x/y) = \log_b x - \log_b y.$

These results can be deduced in a very simple way from the properties of exponentiation. For example, for any positive numbers x and y, numbers p and q can be found such that $x = b^p$ and $y = b^q$, with $\log_b x = p$ and $\log_b y = q$ (for other bases, the values of p and q would change). Thus, $xy = b^p b^q = b^{p+q}$ (from the properties of indices) and, from the definition of the logarithm,

$$\log_b(xy) = \log_b(b^{p+q}) = p + q = \log_b x + \log_b y$$

as given in property (1).

Problem 2.8

For the base e, show that the properties (2) and (3) of logarithms can be deduced by starting from

(a) $x^n = (e^{\ln x})^n = e^{n \ln x}$, for property (2)
(b) $x = e^p, y = e^q$, so that $x/y = e^p/e^q = e^{p-q}$, for property (3).

Worked example

2.6   If $y = 2^{x+2}$, express y in terms of the exponential function.

Solution $y = 2^{x+2} \Rightarrow \ln y = \ln(2^{x+2}) = (x+2)\ln 2 .$

Thus, on using the antilogarithm function and the laws of exponents, $y = e^{(x+2)\ln 2} = e^{x\ln 2} \cdot e^{2\ln 2} \Rightarrow y = 4e^{x\ln 2}$, since $e^{2\ln 2} = e^{\ln 4} = 4$.

Problem 2.9

(a) If $y = f(x) = 3^x = e^{g(x)}$, find $g(x)$, and write down $f(x+h) - f(x)$.

(b) Write $y = (x)^{x^2}$ in the form $y = e^{f(x)}$.

(c) Given that $pH = -\log_{10} a_{H^+}$, show that $a_{H^+} = 10^{-pH} = e^{-pH \ln 10}$.

(d) For a crystalline system consisting of molecular species, which can have two possible orientations at each lattice site ▷, the residual entropy, S, as the temperature approaches 0 K is given by the Boltzmann formula $S = k \ln W$, where W is the total number of arrangements possible, and k is the Boltzmann constant. Assuming there are N molecular species, show that $S = R \ln 2$.

▷

For carbon monoxide, with arrangements like $\cdots$CO OC CO OC $\cdots$ and $\cdots$ CO CO OC OC $\cdots$, the situation is discussed by Berry *et al.*, Section 18.3.

Problem 2.10 demonstrates the use of the exponential function in the Arrhenius equation form that is frequently used to interpret the data associated with the kinetics of a molecular process involving some kind of energy barrier ▷. The same equation appears in different guises: for example, in the study of diffusion phenomena, where the ionic conductivity of solids, such as AgI or LiI, displays phase changes ▷, and in interpreting the electronic conductivity in semiconducting materials, such as Ge. In the latter system, the activation energy for electronic conduction depends upon the band gap ▷. Irrespective of the application, the equation is first transformed to a linear form, through the use of logarithms, as any deviations in the modified data from linearity are much easier to discern. The slope and the intercept on the axis defining the dependent variable are then related to the two parameters in the exponential form of the Arrhenius equation.

▷

Atkins, Section 25.5.

▷

Smart and Moore, Figure 3.7.

▷

Shriver *et al.*, Section 2.10.

Problem 2.10

The Arrhenius equation for the rate constant of a reaction takes the form $k = Ae^{-E/RT}$, where R is the gas constant, E the energy of activation, and T the temperature. Consider the use of this equation for interpreting the following chemical kinetic data that have been obtained for the reaction

$$CH_3I + C_2H_5OH \rightarrow CH_3OC_2H_5 + HI$$

in ethanol ▷, which has a second-order rate constant, k.

(a) Write down the expression relating $\ln k$ to $1/T$.

(b) Use the following data, obtained for the above reaction, to construct a table of $1/T$ and $\ln k$ values, remembering that if $10^5 k = 5.60$ dm^3 mol^{-1} s^{-1} then $k = 5.60 \times 10^{-5}$ dm^3 mol^{-1} s^{-1}.

▷

See Barrow, p. 711.

T/K	$10^5 \, k/\text{dm}^3 \, \text{mol}^{-1} \, \text{s}^{-1}$
273	5.6
279	11.8
285	24.5
291	48.8
303	208.0

(c) Plot a graph of $\ln k$ against $1/T$ for these data, using scales of 4 cm $\equiv 10^{-3} \, \text{K}^{-1} (1/T$ axis) and 1 cm $\equiv 4.0$ ($\ln k$ axis).
(d) Determine a value for A from the intercept on the $\ln k$ axis, and give a rough estimate of the error in its value by determining lines of maximum and minimum slope $\triangleright$.

$\triangleright$
In Chapters 12 and 17 we shall return to the problem of determining the 'best' straight line fit to data.

2.4   Trigonometrical functions

Consider the circle with radius r in Figure 2.4. As the point P moves round the circle in an anticlockwise sense, starting when the angle $\theta = 0$, the length of the horizontal line OQ, defining the x-coordinate of P, starts with a value r; passes through zero, then $-r$, before returning to the initial value of r. The length of the vertical line PQ, defining the y-coordinate of P, starts off at zero and passes through r, zero and $-r$, before returning to zero again as P undertakes a full rotation of $360°$. In both cases the pattern of values repeats itself after every full rotation of P around the circle. The value of x/r defines the value of the cosine function, $\cos \theta$, where θ is the angle $P\widehat{O}Q$; similarly the value of y/r defines the value of the sine function, $\sin \theta$.

For both of these functions, the angle $P\widehat{O}Q$ is the independent variable, and can take on any value (depending how many times P has been rotated around the circle), and the values of each function lie in the interval $[-1, 1]$.

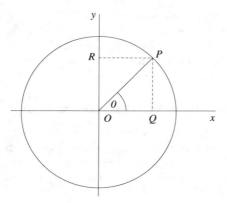

Figure 2.4
Figure used to define the sine and cosine functions

However, unlike the independent variables considered earlier, the magnitude of the angle $P\widehat{O}Q$ can be measured in either in *degrees*($^\circ$), or in *radians*(^{c}). For radian measure, the length of arc on the circle subtended by an angle $P\widehat{O}Q$, divided by r, is defined as θ^c. Thus, for one traversal around the circle, where the arc length is $2\pi r$, 360° is seen to be equivalent to 2π radians; that is $1^\circ \equiv \pi/180$ radians. Hence, the expression $\theta^c = \theta^\circ \cdot \pi/180$ provides a way of interconverting between the two measures of angle.

If the angle is now given the symbol x, then for progressive increments through successive values of 2π, the values obtained for $\cos x$ and $\sin x$ repeat the values obtained in the first region $0 \leq x < 2\pi$: that is, $\cos(x + 2n\pi) = \cos x$ and $\sin(x + 2n\pi) = \sin x$, where $n = 1, 2, \ldots$. Functions displaying this property of cyclically repeating values of the dependent variable are termed *periodic* functions $\triangleright$, and have an important role in chemistry – especially in systems where there is a periodic repetition as seen, for example, in crystals where the repeating pattern defines the unit cell. The tangent, cotangent, secant and cosecant functions are other useful trigonometrical functions that are related to the basic sine and cosine functions:

$$\tan x = \frac{\sin x}{\cos x} \quad \cot x = \frac{\cos x}{\sin x} \quad \sec x = \frac{1}{\cos x} \quad \operatorname{cosec} x = \frac{1}{\sin x}.$$

It should be noted that, while the sine and cosine functions are defined for all real values of the independent variable, none of these related functions has the domain of $\mathbb{R}$, as they all are undefined when either $\cos x$ or $\sin x$ is zero.

Inverse trigonometrical functions

For each of the trigonometrical functions there is an inverse function associated with the reverse mapping from, for example, $\sin x$ back to x. Two notations are used for signifying this inverse function: either a posterior superscript -1 is appended to the function name or 'arc' is added as a prefix. Thus, the inverse function to $\sin x$ is designated either as $\sin^{-1} x$ or arcsin x. A further point worth noting is that, in order to *define* the trigonometric inverse functions $\sin^{-1} x$ and $\cos^{-1} x$ (both of which have the domain $[-1, 1]$), it is necessary to restrict the values of the dependent variable to one period of length 2π to give what are termed the principal values of the inverse function $\triangleright$.

Trigonometrical relations and identities

There is one further notation that is required when dealing with powers of trigonometrical functions. Just as in polynomial expressions, where the notation x^2 is used to signify $x \cdot x$, the analogous notation for $\sin x \cdot \sin x$, for example, is $(\sin x)^2$, which is usually written as $\sin^2 x$.

There are certain key formulae, providing relations between the basic trigonometrical functions, that are important to know, as in many situations their use enables complex expressions to be simplified or other mathematical manipulations to be enacted. Some of the basic formulae are summarized in Appendix 2, where simple proofs of some of the other properties are also given (it is worth working through the steps of the proofs in order to become more familiar with algebraic manipulations).

Problem 2.11

Use the equalities and formulae given in Appendix 2, together with the definitions of the tangent, cotangent, secant and cosecant functions, as appropriate, to prove that

(a) $\cos 2A = 1 - 2\sin^2 A = 2\cos^2 A - 1$, $\sin 2A = 2\sin A \cos A$
(b) $\cos 3A = 4\cos^3 A - 3\cos A$, $\sin 3A = 3\sin A - 4\sin^3 A$
(c) $1 + \tan^2 A = \sec^2 A$, $1 + \cot^2 A = \mathrm{cosec}^2 A$
(d) $\tan(A + B) = \dfrac{\tan A + \tan B}{1 - \tan A \tan B}$, $\tan 2A = \dfrac{2\tan A}{1 - \tan^2 A}$.

2.5   Hyperbolic functions

Linear combinations of the exponential function with exponents x and $-x$ define the *hyperbolic* cosine and sine functions, cosh and sinh respectively $\triangleright$:

$$\cosh x = \frac{e^x + e^{-x}}{2}, \quad \sinh x = \frac{e^x - e^{-x}}{2}.$$

Problem 2.12

From the definitions of $\sinh x$ and $\cosh x$ show that

(a) $\sinh x + \cosh x = e^x$, $\cosh x - \sinh x = e^{-x}$,
    $\ln(\cosh x + \sinh x) = -\ln(\cosh x - \sinh x)$
(b) $\cosh^2 x - \sinh^2 x = 1$, $\cosh^2 x + \sinh^2 x = \cosh 2x$,
    $\sinh 2x = 2\sinh x \cosh x$.

Just as with the trigonometrical functions, quotients of these two new functions can be used to define hyperbolic analogues to $\sec x$, $\operatorname{cosec} x$, $\cot x$, and $\tan x$:

$$\operatorname{sech} x = \frac{1}{\cosh x}, \quad \operatorname{cosech} x = \frac{1}{\sinh x},$$

$$\coth x = \frac{\cosh x}{\sinh x}, \quad \tanh x = \frac{\sinh x}{\cosh x}.$$

The function tanh arises, for example, in studying the adsorption of a gas on a solid surface, and in elucidating magnetic properties of materials.

Problem 2.13

Show that

(a) $\tanh^2 x + \operatorname{sech}^2 x = 1$, $\coth^2 x - \operatorname{cosech}^2 x = 1$

(b) $\tanh x = \dfrac{e^x - e^{-x}}{e^x + e^{-x}} = \dfrac{1 - e^{-2x}}{1 + e^{-2x}} = \dfrac{e^{2x} - 1}{e^{2x} + 1}$.

Problem 2.14

A species with one unpaired electron can exist in two spin states ($\uparrow$, 'spin-up'; $\downarrow$, 'spin-down') of differing energies in the presence of an applied uniform magnetic field. For a field in the direction of the positive z-axis, the 'spin-down' state has the lower energy. If the number of species with spin $\downarrow$ and spin $\uparrow$ is denoted by N_1 and N_2, respectively, and the two energy levels are separated by an energy Δ, then according to the Boltzmann distribution law $N_2 = N_1 e^{-\Delta/kT}$. If the total number of species is N, show that

(a) $N_1 = N/(1 + e^{-\Delta/kT})$; $\quad N_2 = Ne^{-\Delta/kT}/(1 + e^{-\Delta/kT})$

(b) the excess number of species with spin-up over spin-down, X, is given by $X = N\left(\dfrac{1 - e^{-\Delta/kT}}{1 + e^{-\Delta/kT}}\right)$

(c) $X = N \tanh\left(\Delta/(2kT)\right)$, using the result in the Problem 2.13(b).

Inverse hyperbolic functions

The sinh function is defined by $y = \sinh x = (e^x - e^{-x})/2$. The associated inverse function, named $\sinh^{-1}$, is found by making x the subject of the equation $y = \sinh x$; likewise, the $\tanh^{-1}$ function may be defined.

Problem 2.15

(a) Multiply the equation $y = \sinh x = (e^x - e^{-x})/2$ by e^x, and show that $e^{2x} - 2ye^x - 1 = 0$.

(b) Let $w = e^x$, and express w in terms of y.

(c) Show that $x = \ln(y \pm \sqrt{y^2 + 1})$.

(d) By observing that $\sqrt{y^2 + 1} > y$, irrespective of the value for y, explain why $x = \ln(y + \sqrt{y^2 + 1})$ is the only solution, and thus defines the inverse sinh function.

(e) Give the formula for the inverse function using y as the name for the dependent variable.

(f) Follow the procedure used in parts (a) – (d), using $y = \tanh x = (e^x - e^{-x})/(e^x + e^{-x})$ to show that $y = \tanh^{-1} x = \frac{1}{2}\ln\left(\frac{1+x}{1-x}\right)$, with domain $-1 < x < 1$.

Summary: The above discussion of the hyperbolic functions concludes this overview of the idea of function introduced and developed in this chapter. With the aid of examples and problems, we have included most of the functions (and their associated inverse functions) that are appropriate in a chemical context.

The next chapter provides the important tool of a limit for exploring the smoothness of functions, as a necessary exercise in preparing for the calculus, and for later work with functions in chemical applications.

3 Limits, small steps and smoothness

Objectives

This chapter provides

- the basic ideas involved in the mathematical notion of closeness

- an exploration of point properties of functions of a single variable as an aid for investigating their smoothness

- the principles for understanding how the instantaneous rate of change of a function (property) with respect to the independent variable may be defined

- examples in a chemical context where the ideas developed in this chapter may be deployed

The previous chapter described numerous examples of functions that are important within a chemical context. This chapter is concerned with methods for probing the behaviour of such functions, in the close proximity to a point of discontinuity or to a point lying in one of two contiguous intervals of the independent variable. These kinds of technique, applied to functions describing chemical processes associated with changes of concentration, phase, crystal structure, temperature, etc., are important in our approach to differential and integral calculus, which forms the substance of the next three chapters.

3.1 Some examples of limiting processes

We have already seen examples of situations where we need to know what happens to an algebraic expression like $(n - 1)/n$ as n becomes larger and larger. In this particular example the approach adopted was to divide numerator and denominator by n, and then it could be seen that the expression tended to a limiting value of unity. In more complicated examples like

$$\left\{\frac{1}{(n+1)}\cos n\right\} \text{ or } \left\{\frac{3n+\sqrt{n^2-1}}{2n+5}\right\}$$

it is not immediately obvious that both expressions tend towards the well-defined limiting values of 0 and 2, respectively. In another context, it is often necessary to know how particular expressions behave when either one parameter approaches the value of another or one parameter or variable tends to a finite value. We shall see in Chapter 4 that the problem of working out the limiting value of $\sin\theta/\theta$, as θ approaches zero, is non-trivial. A similar and related problem is posed when working out the value of $\sin(a-b)/(a-b)$ as b approaches the value a.

All of these problems and questions are resolved by making use of the mathematical idea of a limiting process. The formal treatment of such processes requires some delicate arguments to be deployed which are somewhat abstract in nature; we adopt a more pragmatic approach in order to provide a working set of tools for applications in this and later chapters.

3.2   Defining the limiting process

Consider the function $y = f(x)$, with some specified domain (a meaningful set of x values, the nature of which is left unspecified at the moment). If $f(x)$ approaches the value m as x becomes closer and closer to the value a, then we say that m is the limit of $f(x)$ as x approaches a:

$$\lim_{x\to a} f(x) = m \tag{3.1}$$

It is important to note that $x \to a$ is a symbolism to remind us that x approaches a as close as we like, but may never actually take the value a. For those functions, where the domain is such that a can be approached from both smaller and larger values of x, the limit exists only if the same finite value of m is obtained from either direction of approach: that is

$$\lim_{x\to a^+} f(x) = \lim_{x\to a^-} f(x) = m\,, \tag{3.2}$$

where a^+ and a^- correspond to values of x larger or smaller than a, respectively.

Functions of an integer variable

Examples and problems, illustrating the application of Equation (3.1) to functions with domain $n = 0, 1, 2, \ldots$, are first presented in order to explore further the technique for deriving the limiting values of such functions as n tends to an indefinitely large number; that is $n \to \infty$.

Worked example

▷
A subset of $\mathbb{I}$.

3.1   Given the function $y = 2^{-n}$, with domain $\{n = 0, 1, 2, 3, 4, \ldots\}$ ▷,

(a) determine the limiting value of the sequence of y values
$$1, \frac{1}{2}, \frac{1}{4}, \frac{1}{8}, \ldots \frac{1}{2^n}, \text{ as } n \text{ becomes indefinitely large;}$$

(b) deduce the form of $y = 2^{-n}$, giving its domain.

Solution   (a) The domain of the function $y = f(n) = 2^{-n}$ is the set $\{n = 0, 1, 2, 3, 4, \ldots\}$ and, in the limit as n becomes indefinitely large, 2^{-n} becomes progressively closer to zero, i.e.,
$$\lim_{n \to \infty} 2^{-n} = \lim_{n \to \infty} \frac{1}{2^n} = 0.$$

▷
A subset of the rational numbers.

(b) The domain of the inverse function is $\{y = 1, 2^{-1}, 2^{-2}, 2^{-3} \ldots 2^{-n}, \ldots\}$, ▷ and its form is found by making n the subject of the equation defining y:
$$y = 2^{-n} \Rightarrow -n = \log_2 y \Rightarrow n = -\log_2 y.$$

Worked example

3.2   Find the limit of the sequence generated by
$$y = f(n) = \left\{ \frac{\sqrt{n^2 - 1}}{3n} \right\}.$$

Solution   For a given n,
$$\left\{ \frac{\sqrt{n^2 - 1}}{3n} \right\} = \left\{ \frac{1n\sqrt{n^2 - 1}}{3} \right\} = \left\{ \frac{\sqrt{1n^2(n^2 - 1)}}{3} \right\} = \left\{ \frac{\sqrt{1 - 1/n^2}}{3} \right\},$$

where numerator and denominator have both been divided by n. Since, for large n, $1/n^2$ can be ignored in relation to 1, $y(n)$ tends to $1/3$.

Problem 3.1

Determine which of the following sequences $\{f(n)\}$ (n in the domain of f) converge, and give their respective limits as n becomes indefinitely large.

(a) $y = f(n) = \dfrac{3n + \sqrt{n^2 - 1}}{2n + 5}$, domain$\{1, 2, 3, \ldots\}$

(b) $y = f(n) = \sin n$, domain$\{0, 1, 2, \ldots\}$

(c) $y = f(n) = \dfrac{1}{n + 1} \cdot \sin n$, domain$\{0, 1, 2, \ldots\}$.

If the domains of the function $f(n)$ in (a) – (c) are extended to include negative integers, give the limiting values (where they exist) of $f(n)$ as n tends to $-\infty$.

Functions of a real variable

We now turn to the application of Equations (3.1), (3.2) to functions of a real variable, $y = f(x)$. Of especial interest is the behaviour of $f(x)$ as x approaches a chosen finite value, in order to explore the point behaviour of the function. It may also be of interest to examine the asymptotic values of $f(x)$, in the limits as $x \to \pm\infty$.

Consider, for example, the function

$$y = f(x) = \frac{x^2 - x - 2}{x - 2}, \tag{3.3}$$

▷

The value of $f(2) = 0/0$ and is undefined.

where it is important to know how $f(x)$ behaves in the vicinity of $x = 2$, because this value of x is clearly not in the domain of the function ▷. In this kind of situation, it is often helpful to write $x = 2 + \delta$, and consider what happens as δ approaches 0 (this is the same as considering x approaching 2):

$$\lim_{\delta \to 0} \frac{(2 + \delta)^2 - (2 + \delta) - 2}{2 + \delta - 2} = \lim_{\delta \to 0} \frac{4 + 4\delta + \delta^2 - 4 - \delta}{\delta}$$

$$= \lim_{\delta \to 0} \frac{3\delta + \delta^2}{\delta} = \lim_{\delta \to 0} \frac{3 + \delta}{1} = 3.$$

Notice that,

▷

δ is negative on approaching 2 from values of $x < 2$, and positive for $x > 2$.

- in the penultimate step, division by δ is permitted as its value can be made as small as we like, but never zero;

- irrespective of the sign of δ, the same limiting value of 3 is obtained, thus confirming the requirements of Equation (3.2) ▷.

Testing for continuity

For the limit in Equation (3.1) to exist, for a given arbitrary function $f(x)$, the two different 'half-limits', $\lim_{x \to a^+} f(x)$ and $\lim_{x \to a^-} f(x)$ (Equation (3.2)) must yield the same finite value, m. Now, although the limit may exist at the

▷

$f(2)$ is undefined in Equation 3.3.

▷

As observed in a phase change.

point a, the value of the function there, $f(a)$, may be undefined ▷, infinite or finite; for a finite value of $f(a)$, the function is *continuous* at a only if $f(a) = m$; otherwise it displays a finite discontinuity ▷. If $f(a)$ is infinite, then the function exhibits an infinite discontinuity at the point with $x = a$. Continuous functions are smooth in the sense that if graphs are drawn of them then it is not necessary to lift the pencil off the paper. As already noted above, and in the previous chapter, we do meet functions displaying discontinuities (especially in a thermodynamic context), and this is why we need to detect any particular point(s) where there might be a discontinuity.

Problem 3.2

Evaluate $\lim_{x \to a} f(x)$ for the values of a and the functions defined below. Comment on the continuity of each of the functions, and identify any points of discontinuity.

(a) $\lim_{x \to 2} \dfrac{1}{(x^2 - 2)}$, $\lim_{x \to 3} (x^2 - x + 1)$, $\lim_{x \to 2} \dfrac{x}{x - 3}$.

(b) $\lim_{x \to 2} \dfrac{x^2 - 4}{x - 2}$, $\lim_{x \to 2} \left(\dfrac{x^2 - 4}{x - 2} \right) \cos \left(\dfrac{\pi x}{4} \right)$, $\lim_{x \to \infty} \dfrac{x}{2x + 1}$,

$\lim_{x \to \infty} \dfrac{x^n}{(x^2 + a^2)^{n/2}}$.

3.3   Some examples in the use of limits

Problem 3.3

The rate of decomposition, v, of phosphine on tungsten is given by

$$v = \frac{kKp}{1 + Kp}$$

▷

Atkins, p. 998.

▷

<< and >> mean very much smaller and very much greater, respectively.

where p is the pressure of phosphine, k is the rate constant, and K is the ratio of the rate constants for adsorption and desorption ▷. Notice that K carries units of inverse pressure. Determine the orders of the decomposition reaction when p is such that (a) $Kp << 1$, (b) $Kp >> 1$ ▷.

▷
Alberty and Silbey,
p. 652.

Problem 3.4

Bodenstein observed that the rate of the reaction of gaseous bromine with hydrogen to produce hydrogen bromide could be described empirically ▷ by

$$\frac{k[H_2][Br_2]^{12}}{1 + k'[HBr]/[Br_2]} \, .$$

Determine how the initial rate depends upon the concentration of Br_2 at the start of the reaction when (a) $k'[HBr]/[Br_2] \ll 1$, (b) sufficient HBr is added to ensure that $k'[HBr]/[Br_2] \gg 1$.

Problem 3.5

For the function defined by the prescription

$$y = f(x) = \begin{cases} x+1, & \text{if } x \leq 2 \\ -x+5, & \text{if } x > 2 \end{cases}$$

(a) evaluate $\lim_{x \to 2^+} f(x)$, $\lim_{x \to 2^-} f(x)$, and $f(2)$, and show that $f(x)$ is continuous at $x = 2$;

(b) sketch the form of the function for $1 < x < 3$.

Worked example

3.3 For the $f(x)$ defined in Problem 2.9(a), show that

$$\lim_{h \to 0} \left(\frac{f(x+h) - f(x)}{h} \right) = \ln 3 \cdot e^{x \ln 3} \, .$$

Solution From Problem 2.9(a) we have $y = e^{x \ln 3} = e^{ax}$, where $a = \ln 3$. Thus

$$f(x+h) - f(x) = e^{a(x+h)} - e^{ax} = e^{ax}(e^{ah} - 1) \, ,$$

and

$$\lim_{h \to 0} \left(\frac{e^{ax}(e^{ah} - 1)}{h} \right) = e^{ax} \cdot \lim_{h \to 0} \left(\frac{e^{ah} - 1}{h} \right) \, .$$

We observe that, as $h \to 0$, both denominator and numerator $\to 0$ (but at different rates). $0/0$ is obviously an undefined quantity, so we must handle the limiting process with more care. It was shown earlier in Chapter 2 (Equation (2.1)) that e^x can be represented in the form of the series

$$e^x = 1 + x + \frac{x^2}{2!} + \cdots$$

▷
Small x usually means
$x \ll 1$.

for small x ▷. Hence substituting ah for x in this expression yields
$e^{ah} = 1 + ah + \frac{1}{2}(ah)^2 + \cdots$

$$\Rightarrow \lim_{h \to 0} \left(\frac{f(x+h) - f(x)}{h} \right)$$

$$= e^{ax} \cdot \lim_{h \to 0} \left(\frac{ah + 12(ah)^2 + \cdots}{h} \right) = e^{ax} \cdot \lim_{h \to 0} \left(a + \frac{1}{2} ha^2 + \cdots \right)$$

$$= \ln 3 \, e^{x \ln 3},$$

▷
Remember that
$a = \ln 3$.

as all higher terms go to zero ▷.

Problem 3.6

▷
Atkins, p. 362.

(a) The formula for the Planck distribution of energy per unit volume per unit wavelength, ρ, emitted by a black body is given by ▷:

$$\rho = \frac{8\pi hc}{\lambda^5} \left\{ \frac{1}{e^{hc/\lambda kT} - 1} \right\}.$$

Show that, for sufficiently large λ, when the expansion for e^x given in Example 3.3 can be used,

$$\rho \to \frac{8\pi kT}{\lambda^4}.$$

▷
Atkins, p. 364.

(b) The Einstein model for the molar heat capacity at constant volume, C_v, of a solid yields the formula ▷:

$$C_v = 3R \left(\frac{h\nu}{kT} \right)^2 \left\{ \frac{e^{h\nu 2kT}}{e^{h\nu kT} - 1} \right\}^2.$$

Show that at very high values of T, C_v tends to the limit $3R$.

(c) For the sequential first-order chemical reactions $A \xrightarrow{k_1} B \xrightarrow{k_2} C$, the concentration of B at time t is given by

$$b = \frac{\alpha k_1}{k_2 - k_1} \left(e^{-k_1 t} - e^{-k_2 t} \right),$$

where α is the concentration of A at time $t = 0$. In the situation where $k_1 = k_2$, it is necessary to evaluate the limit $\lim_{k_2 \to k_1} b$. First substitute $k_2 = k_1 + \delta$, and then use the results that
(i) $\lim_{k_2 \to k_1} b = \lim_{\delta \to 0} b$ (ii) $e^{-\delta t} = 1 - (\delta t) + (\delta t)^2 / 2 \cdots$,
and thus derive the expression for b as a function of t in the situation where $k_1 = k_2$.

Summary: This excursion into a number of aspects of limiting processes, involving smoothness and asymptotic (limiting) forms of functions, provides the basis for both differential and integral calculus, which are developed in Chapters 4–6. Of particular importance is the recognition that it is necessary to have the confidence to handle extrapolation of formulae arising from the chemical modelling of reality to cover the cases where the variable(s) in the formulae take on extremely large or small values.

4 Rates of change and differentiation

Objectives

This chapter

- shows how the idea of limit can be used to move from the average change of a function (property) over a given interval to the instantaneous rate of change at a particular value of the independent variable to yield the derivative of the function

- extends the notion of derivative to sums, products and quotients of functions

- provides the techniques used for differentiating functions which can be expressed in terms of other functions

▷

There are, of course, some chemical reactions that take place under the influence of a light source or following impact by an excited species.

Chemistry is concerned largely with the study of systems whose chemical or physical properties evolve with time, concentration, or some other variable, in a characteristic way: thus, typically, a chemical reaction occurs at a rate which depends upon the change in one or more concentrations of the reacting species with time. In systems where there is usually no chemical reaction, as in the various branches of spectroscopy applied to species in their ground electronic states, the main point of interest lies in the response of the atom, molecule, solid, etc., to an electromagnetic field whose strength fluctuates with a characteristic wavelength ▷ (the function describing the field is periodic, in the sense that the fluctuation repeats itself after every characteristic wavelength). In all these, and other, situations it is important to understand the techniques required for describing rates of change in a mathematical sense. For this purpose it is easiest to start with a function of a single variable which may, for example, describe how concentration varies with time; using the techniques of calculus as described below, it is then possible to discover how this variation may be described mathematically.

4.1   Defining rate of change

Average rate of change

Consider the graphical representation of the function defined by $y = f(x)$ (the full line in Figure 4.1). The change in the value of the function between x_0 and $x_0 + h$ is given by $\Delta y = f(x_0 + h) - f(x_0)$: thus the average rate of change of $f(x)$ with x over the interval of width h ▷ is given by

▷
Sometimes written as Δx, the increment in x.

$$\frac{f(x_0 + h) - f(x_0)}{h} = \frac{\Delta y}{h} = \frac{QR}{PR},$$

which is the slope of the secant PQ.

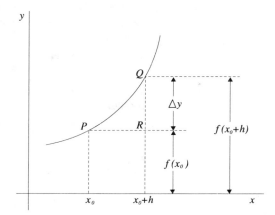

Figure 4.1
Construction to define the average rate of change of a function in an interval

Instantaneous rate of change

▷
If this limit exists.

We can determine the rate of change of $f(x)$ with x over progressively smaller and smaller intervals until, in the limit as h approaches zero ▷, we have the instantaneous rate of change of $f(x)$ with x at x_0: a quantity which is termed the *derivative* of $f(x)$ and, at x_0, is defined by

$$\left(\frac{dy}{dx}\right)_{x=x_0} = \lim_{h \to 0} \left(\frac{f(x_0 + h) - f(x_0)}{h}\right).$$

As already noted, this process is meaningful only if the limit exists and, in such a case, it makes no difference whether x_0 is approached from greater or smaller values of x (h is positive or negative, respectively). The value of this limit describes the rate of change of $f(x)$ with respect to x at the point x_0, and is given the symbol $\left(\dfrac{dy}{dx}\right)_{x=x_0}$. Alternative symbols for this derivative (which is, in general, a new function of x) are

$$\left(\frac{df(x)}{dx}\right)_{x=x_0}, \; f'(x_0) \quad \text{or} \quad f^{(1)}(x_0). \tag{4.1}$$

From now on, unless it is necessary to specify the value of x at which the differentiation is carried out, explicit reference to x_0 is dropped, and we write

$$\frac{dy}{dx} = \frac{df(x)}{dx} \equiv f'(x) = \lim_{h \to 0}\left(\frac{f(x+h) - f(x)}{h}\right) \equiv \frac{d}{dx} f(x) \equiv \hat{D}f(x),$$
$$\tag{4.2}$$

where x is the value of the independent variable in the domain of the function. The five ways of designating the derivative of $f(x)$ in Equation (4.2) are all in use, and the common one, dy/dx, should be read as one symbol – not a quotient. For this reason, the notation $f'(x)$ is preferred; however, expressing the derivative in the form $\hat{D}f(x)$ is also useful because it indicates clearly that the derivative function is obtained from $f(x)$ through the operation of differentiation. The values of x for which the derivative function is defined (the domain) may differ from those for $f(x)$.

Geometrically, $f'(x_0)$ corresponds to the slope of the straight line which touches the curve $y = f(x)$ at the point (x_0, y_0) ▷.

▷
This straight line is the tangent to the curve at the point (x_0, y_0).

4.2   Differentiation of some standard functions

In principle, for any function, $f(x)$, the derivative function can be obtained by evaluating the limit in Equation (4.2); however, this is very tedious, and we shall follow the procedure of using this method only for the standard functions $y = x^n$, $y = e^x$, $y = \sin x$, $y = \cos x$ and $y = \ln x$ through the use of structured problems. Derivatives of other related functions, which can be expressed in terms of products, sums and quotients (or any combination of these) of the standard functions, can then be found by successive application of the rules for differentiating a sum, product or quotient of functions – as seen in the next section. In other cases, where it is not immediately obvious how to proceed, the strategy of introducing a temporary change of variable is often invaluable: again, this method is discussed in a later section. Practice is the key to fluency in the calculus! We do not advocate the derivative analogue of 'The Daily Integral' routine of Matthews ▷, but it is important to learn the basic strategies and to use lists of standard derivatives, some of which are derived in this chapter and summarized along with other results in Appendix 3. When in real need, it is best to consult the much more extensive compilations of derivatives and other mathematical data ▷.

▷
For example p. 183.
▷
For example Dwight, Abramowitz and Stegun or Gradshteyn and Ryzhik.

Differentiation of x^n

Worked example

4.1 Find the derivative of $y = f(x) = x^2$, and give the domain of the derivative function.

Solution
$$\lim_{h\to 0}\left(\frac{f(x+h)-f(x)}{h}\right) = \lim_{h\to 0}\left(\frac{(x+h)^2 - x^2}{h}\right) = \lim_{h\to 0}\left(\frac{2xh + h^2}{h}\right)$$

$$= \lim_{h\to 0}(2x + h) = 2x.$$

▷
That is, they both have the domain $\mathbb{R}$.

In this example, $f(x)$ and $f'(x)$ are both defined for all x ▷.

Problem 4.1

(a) Use Equation (4.2) to demonstrate that the derivative of the constant function $y = f(x) = c$ is zero.

(b) From the result of Problem 2.3(c), deduce that the derivative of the function $g(x) = 1/x$ is given by $g'(x) = -1/x^2$.

▷
See Chapter 1 for the expansion of $(1 + x)^n$, where x is replaced by h/x.

(c) Use the definition of derivative given in Equation (4.2), together with the result ▷ $(x+h)^n = x^n(1 + h/x)^n = x^n + nx^{n-1}h + n(n-1)x^{n-2}h^2/2 + \cdots$ to find the derivative of $y = x^n$. Give the derivatives for $n = 2$, $n = -3$ and $n = 5/2$.

Differentiation of $\sin x$ and $\cos x$

The differentiation of the trigonometrical functions requires a new result. Consider, for example, $y = f(x) = \sin x$. It follows directly from the the expansion $\sin(x + h) = \sin x \cos h + \cos x \sin h$ that

$$\frac{f(x+h)-f(x)}{h} = \frac{\sin(x+h)-\sin x}{h}$$

$$= \frac{\sin x \cos h}{h} + \frac{\cos x \sin h}{h} - \frac{\sin x}{h}$$

$$= \sin x \frac{(\cos h - 1)}{h} + \cos x \frac{\sin h}{h}$$

$$= -\sin x \cdot \frac{\sin(h/2)}{h/2}\cdot \sin(h/2) + \cos x \cdot \frac{\sin h}{h} \quad (4.3)$$

▷
See Problem 2.14(a).

after using the identity $1 - \cos h = 2\sin^2(h/2)$ ▷.

We are now faced with a problem because, on taking the limit as $h \to 0$, the two terms $\dfrac{\sin(h/2)}{h/2}$ and $\dfrac{\sin h}{h}$ apparently tend to 0/0 in the limiting process as $h \to 0$. This is, in fact, not the case and a short interlude now follows to show that the limit of the sine of an angle divided by the angle tends to unity as the angle tends to zero.

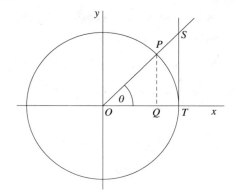

Figure 4.2
The geometrical
construction, involving a
circle of radius r, for
evaluating $\sin\theta/\theta$ in the
limit as $\theta \to 0$

An important limit

In Figure 4.2, three regions can be discerned: the right-angled triangular regions OPQ, OST and the sector OPT (θ is in radian measure). The areas of the two triangular regions, A_1 and A_2, respectively, are found by using the simple rule of base times height divided by two for the area of a triangle. The area of the sector OPT is found by observing that the ratio of the length of a circular arc to that of the full circumference is equal to the ratio of the area of the corresponding sector, A_3, to the area of the circle: that is

$$\frac{r\theta}{2\pi r} = \frac{A_3}{\pi r^2} \Rightarrow A_3 = \frac{r^2\theta}{2}.$$

For a circle of unit radius, we see that $A_1 = \frac{1}{2}\cos\theta\sin\theta$, $A_2 = \frac{1}{2}\tan\theta$, $A_3 = \theta/2$.

Since $A_2 > A_3 > A_1$, we can write

$$\frac{1}{2}\tan\theta > \frac{\theta}{2} > \frac{1}{2}\cos\theta\sin\theta \Rightarrow \sec\theta > \frac{\theta}{\sin\theta} > \cos\theta,$$

$\triangleright$
Since θ is small and
positive, $\sin\theta$ is
positive, and hence the
sense of the inequalities
is left unchanged by this
division.

after dividing through by $\sin\theta$ $\triangleright$. This sequence of inequalities can now be inverted as all the factors are positive:

$$\cos\theta < \frac{\sin\theta}{\theta} < \sec\theta. \tag{4.4}$$

Now for very small positive values of θ, $\cos\theta$ is positive and slightly less than unity, whilst $\sec\theta = 1/\cos\theta$ is slightly greater than unity.

$\triangleright$

$\approx$ means 'is closely
approximated by'.

In Chapter 7, it is seen that the respective departures of these values from unity can be expressed as $-\theta^2/2$ and $\theta^2/2$, so that, in the limit of very small positive θ, $\cos\theta \approx 1 - \theta^2/2$ and $\sec\theta \approx 1 + \theta^2/2$ $\triangleright$; and Equation (4.4) becomes

$$1 - \theta^2/2 < \frac{\sin\theta}{\theta} < 1 + \theta^2/2.$$

In the limit as $\theta \to 0$, it is seen that $\sin\theta/\theta$ tends to 1, as its value is squeezed between two numbers $1 - \theta^2/2$ and $1 + \theta^2/2$ which both approach 1 in the

▷
The limit of a product is
the product of the limits.

limit. It therefore follows that ▷

$$\lim_{\theta \to 0} \frac{\sin \theta}{\theta} = 1 . \tag{4.5}$$

Thus, using this result in Equation (4.3), first with $\theta = h/2$ and then with $\theta = h$, it follows that

$$f'(x) = -\sin x \lim_{h \to 0} \frac{\sin(h/2)}{h/2} \cdot \sin(h/2) + \cos x \lim_{h \to 0} \frac{\sin h}{h}$$

$$= \sin x \cdot 1 \cdot 0 + \cos x = \cos x .$$

Problem 4.2

For $f(x) = \cos x$, demonstrate that

$$\frac{f(x+h) - f(x)}{h} = -2 \cos x \cdot \frac{\sin^2(h/2)}{h} - \sin x \cdot \frac{\sin h}{h} ,$$

and use Equation (4.5), with θ replaced by $h/2$ and then h, to show that $f'(x) = -\sin x$.

Differentiating the exponential and logarithm functions

The derivatives of these important functions are formulated in the next problem.

Problem 4.3

(a) Re-read Example 3.3, and show that the derivative of e^{ax} is given by

$$\frac{d}{dx} e^{ax} = ae^{ax}.$$

(b) If g is defined as $\dfrac{f(x+h) - f(x)}{h}$ then, for $f(x) = \ln x$, show that

$$g = \frac{\ln(1 + h/x)}{h} \Rightarrow gh = \ln(1 + h/x) \Rightarrow e^{gh} = (1 + h/x) \, ▷.$$

1. Use the expansion for e^{gh} given in Example 3.3 to show that

$$gh + (gh)^2 + \cdots = h/x , \tag{4.6}$$

2. Divide through Equation (4.6) by h, and demonstrate that

$$\lim_{h \to 0} g = 1/x .$$

▷
Note that
$\lim_{h \to 0} g = f'(x)$.

4.3   Functions with discontinuities

The method described above for obtaining the derivative function presents difficulties if different limiting values are obtained when approaching zero from positive or negative values of h. In these situations, the function is exhibiting a discontinuity at the point in question, and it is not differentiable at such a point. In practice, the formula or prescription defining the function displays very clearly where such points are likely to occur: for example, with $y = \tan x$, at odd multiples of $\pi/2$. In a chemical context, where there are phase changes, discontinuities preclude the differentiation of the function describing a particular physical property at appropriate transition temperatures. For example, the variation of entropy, S, with temperature, given schematically in Figure 4.3, shows discontinuities at temperatures where there is a change of physical phase: here at T_s, where there is a change of phase in the solid ▷, and at the melting point (T_m) and boiling point (T_b), respectively. At these transition temperatures, the derivative dS/dT is not defined, and the entropy function displays finite discontinuities.

▷

As seen, for example, in solid n-butane.

Some functions, for example the tangent function, $\tan x$, display infinite discontinuities at points where $\cos x$ is zero (for $x = (2n + 1)\pi/2$).

Figure 4.3
The variation of entropy with temperature, as an example of a function with discontinuities at T_s, T_m, T_b. T_s is the temperature where a phase change occurs in the solid, and T_m, T_b are the melting and boiling points, respectively

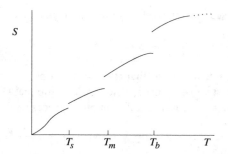

4.4   Basic rules for differentiation

Sums, products and quotients of functions

▷

See, for example, Salas and Hille, Section 3.2.

The basic way of defining a derivative described above can be used to determine the form of the derived function for a sum, product and quotient of functions. The proofs are relatively straightforward, and can be found in all textbooks on calculus ▷. The results are stated here without proof:

$$\frac{d}{dx}(f(x) + g(x)) = f'(x) + g'(x) \tag{4.7}$$

$$\frac{d}{dx}(f(x)g(x)) = f'(x)g(x) + f(x)g'(x) \tag{4.8}$$

$$\frac{d}{dx}\left(\frac{f(x)}{g(x)}\right) = \frac{g(x)f'(x) - f(x)g'(x)}{[g(x)]^2} \tag{4.9}$$

These rules enable us to extend the list of simple derivatives given in Appendix 3.

Worked example

4.2   Find the derivatives of the functions $y = x^2 + 18$ and $y = 3x^2$.

Solution   Equation (4.2) can be used to obtain both derivatives from first principles; however, Equations (4.7) and (4.8) provide simple routes to the required results.

For the first function, $f(x)$ and $g(x)$ in Equation (4.7) may be identified with x^2 and 18, repectively, so that the results in Problem 4.1(a) and Example 4.1 can be used directly to yield

$$\frac{d}{dx}(x^2 + 18) = 2x + 0 = 2x.$$

For the second function, $f(x) = 3$ and $g(x) = x^2$, and

$$\frac{d}{dx}3x^2 = 0 \cdot x^2 + 3 \cdot 2x = 6x.$$

In this case, it is clear that the result can be obtained directly by multiplying the derivative of x^2 by 3, since the (multiplicative) constant factor 3 plays a subordinate role in the differentiation process.

Problem 4.4

Identify possible functions $f(x)$ and $g(x)$, and use the above formulae to differentiate the following:

(a) $x + \sin x$ (b) $(3x - 9)(x^2 + 18)$ (c) e^{2x}
(d) $x^2/(x - 1)$ (e) $\cos^2 x$ (f) $x \ln x$
(g) $\tan x$ (h) $xe^{2x} \ln x$ (i) $\tan^2 x$
(j) $\sinh x$ (k) $\cosh x$
(l) $(\cos x - \sin x)/(\cos x + \sin x)$.

In a chemical context the need to work with derivatives becomes apparent when dealing with functions describing how one property changes with respect to another property which is acting as the independent variable: thus, in kinetics, the derivative $d[A]/dt$ describes the rate of change of concentration of the species A with time. Similarly, in many instances, chemical data are fitted to a straight line plot, where the function is $y = mx + c$, where m is the slope of the line, and c the intercept on the y-axis; in these cases, it is clear that $dy/dx = m$, i.e., the derivative function – here a constant – yields the slope of the straight line.

The chain rule

As already noted above, the tools we have developed are really only useful for dealing with the differentiation of fairly simple functions. Quite frequently we are faced with the problem of differentiating functions like

$$y = \cos(3x^2 + 1) \text{ or } y = (1 + 3x)^4,$$

for example. We know how to differentiate the simple functions $f(x) = \cos x$ and $g(x) = 3x^2 + 1$, but it is not at all clear how this knowledge can be deployed to find the derivative of what is a function of a function $\triangleright$; likewise the second example can be viewed as $f(g(x))$, where $f(x) = x^4$ and $g(x) = 1 + 3x$. The question remains, how do we find the derivative of $f(g(x))$? The trick is to see that $g(x)$ can be treated as a new, or intermediate, variable u.

$\triangleright$
The first function given can be written as $f(g(x))$, where $g(x)$ plays the role of the independent variable.

Let

$$y = f(g(x)) = \cos(3x^2 + 1)$$

and

$$u = g(x) = (3x^2 + 1)$$

so that we can write $y = f(u)$. The rule for differentiating y is quite simple:

$$\frac{dy}{dx} = \frac{d}{dx}f(u) = \frac{d}{du}f(u) \cdot \frac{d}{dx}g(x),$$

which may be written as

$$\frac{dy}{dx} = \frac{dy}{du} \cdot \frac{du}{dx}.$$

Thus, for the example above,

$$\frac{dy}{du} = -\sin u, \quad \frac{du}{dx} = 6x.$$

Therefore

$$\frac{dy}{dx} = -\sin u \times 6x = -6x \sin(3x^2 + 1).$$

The last step is especially important as, at the end of the analysis, the intermediate variable should be eliminated in favour of x by substituting the function $g(x)$ wherever u appears in the expression for the derivative. Only when the result would be very cumbersome if this were done should an exception be made to this rule.

Problem 4.5

Use the chain rule to find the derivatives of

(a) $y = e^{\sin x}$, $y = e^{h(x)}$, $y = 4\sin(x/2)$

(b) $y = \ln(\ln x)$, $y = (1 + 3x)^4$, $y = \cos(3x^2 + 1)$.

Problem 4.6

Alberty and Silbey,
Example 14.6.

In the ground vibrational state of a diatomic molecule like $^1\text{H}^{35}\text{Cl}$, the fractional occupation of the rotational state with quantum number J at temperature T is given by the function $\triangleright$

$$y = h(J) = \frac{(2J + 1)\,e^{-J(J+1)\Theta_r/T}}{q_r},$$

where $\Theta_r = hcB_v/k$, $q_r = T/\Theta_r$, and $h(J)$ is approximated by its envelope function (the continuous function passing through all the points $(J, h(J))$). Determine the derivative of $h(J)$ with respect to J.

4.5   Higher-order derivatives

If we are given $y = f(x)$, then $dy/dx = f'(x)$, the derivative function, is in general another function of x, $h(x)$, say. This function may then be differentiated again to yield the second derivative of $f(x)$, which we write as

$$f''(x), \quad f^{(2)}(x), \quad \frac{d^2 f(x)}{dx^2}, \quad \text{or } \frac{d^2}{dx^2} f(x).$$

This process may be repeated to define an nth order derivative, $f^{(n)}(x)$, provided that each of the earlier derivatives is differentiable: that is, the respective limits, given in Equation (4.2), exist for each successive derivative function. The notation introduced here of superscripting the function name with (n) is especially useful in obtaining formulae for a derivative of arbitrary order; $f^{(0)}(x)$ is then used to signify $f(x)$.

Worked example

4.3 For $f(x) = (x+1)^3$, find d^2y/dx^2.

Solution The first step is to find the derivative dy/dx. This is best accomplished by using the chain rule as described above. Let $u = (x+1)$, then

$$\frac{dy}{dx} = \frac{dy}{du} \cdot \frac{du}{dx} = 3u^2 \cdot 1 = 3(x+1)^2 = h(x) = w, \text{ (say)}.$$

But dw/dx can be found by using the chain rule again, letting $u = (x+1)$:

$$\frac{dw}{dx} = \frac{dw}{du} \cdot \frac{du}{dx} = 3 \cdot 2 \cdot (x+1).$$

Therefore

$$\frac{dw}{dx} = \frac{d}{dx} h(x) = \frac{d^2y}{dx^2} = \frac{d^2}{dx^2} f(x) = 3 \cdot 2 \cdot (x+1) = 6(x+1).$$

Worked example

4.4 Find the first, second, third and nth-order derivatives of the function $y = f(x) = (1+x)^{-1}$.

▷
The domain is
$\mathbb{R} - \{-1\}$.

Solution Notice that the domain of $f(x)$ is defined by excluding the point $x = -1$ from the set of real numbers ▷.

$$f^{(1)}(x) = -1(1+x)^{-2}, \text{ using the chain rule with } u = 1+x$$
$$f^{(2)}(x) = -1 \cdot -2(1+x)^{-3} = (-1)^2 \, 2! \, (1+x)^{-3}$$
$$f^{(3)}(x) = -1 \cdot -2 \cdot -3(1+x)^{-4} = (-1)^3 \, 3! \, (1+x)^{-4}.$$

The pattern is now apparent for guessing the form of $f^{(n)}(x)$: the phase factor -1 is raised to the power n; the power of $(1+x)$ is the negative of the number corresponding to one more than the number of differentiations, and the additional factor of $n!$ is produced. Thus,

$$f^{(n)}(x) = -1 \cdot -2 \cdot -3 \cdot \ldots \cdot -n(1+x)^{-(n+1)}$$
$$= (-1)^n \, n! \, (1+x)^{-(n+1)}, \quad (n = 0, 1, 2, \ldots). \tag{4.10}$$

Problem 4.7

Find the nth derivative of each of the following functions, giving the values of x where the derivative is not defined. It may be necessary to differentiate at least four times before the pattern becomes apparent.

(a) $f(x) = \ln(1 + x)$
(b) $f(x) = e^{ax}$
(c) $f(x) = \sin x$ (see hint (1) below)
(d) $f(x) = (1 + x)^{1/2}$ (see hint (2) below).

Hints :

1. The function and its derivative functions repeat themselves after every four differentiations. The general formula for the nth derivative is found most easily by noting that

$$f^{(2)}(x) = -\sin x = \cos(x + \pi/2), \text{ and}$$
$$f^{(3)}(x) = -\cos x = \cos(x + 2 \cdot \pi/2), \text{ etc.}$$

2. The product $1 \cdot 3 \cdot 5 \cdots (2n - 3)$ may be written as

$$\frac{1 \cdot 3 \cdot 5 \cdots (2n - 3) \times 2 \cdot 4 \cdots (2n - 4)}{2 \cdot 4 \cdots (2n - 4)} = \frac{(2n - 3)!}{2^{n-2}(n - 2)!}.$$

See Atkins, p. A8.

Problem 4.8

The modelling of the ionic atmosphere in solutions is described in the Debye–Hückel theory by means of a suitable potential, ϕ_i, which depends upon a parameter termed the Debye length, r_D ▷. This potential satisfies the equation

$$\frac{1}{r^2}\frac{d}{dr}\left(r^2 \frac{d\phi_i}{dr}\right) = \frac{2\rho F^2 I \phi_i}{\epsilon RT},$$

where ρ is the density of the solution and the other parameters are either fundamental constants or determined by the conditions of the experiment.

By using ϕ_i in the form

$$\phi_i = \frac{Z_i}{r} e^{-r/r_D},$$

where Z_i just depends upon the charge of the ith ion and the permittivity of the medium, find its derivative with respect to r, and hence verify that

$$r_D = \left(\frac{\epsilon RT}{2\rho F^2 I}\right)^{1/2}.$$

4.6   Maxima and minima

Values of x for which $f'(x) = 0$ are termed *turning points*. A turning point may indicate the presence of either a *maximum* or a *minimum* of $f(x)$ within a certain interval of x values; the *global* maximum or minimum value of the function may occur either at an end-point of the interval of x values, or correspond to one of the turning points.

Since $f'(x)$ is the slope of the tangent at the point (x, y), the tangent line has zero slope at a turning point. It is not possible, therefore, without further information, to discern whether a given turning point corresponds to a local minimum or maximum value of $f(x)$. The characterization of a turning point involves calculating the value of the second derivative $\triangleright$ at the turning point. Thus, on passing through a minimum (maximum) from smaller to larger values of x about x_0, the slope changes from negative (positive), through zero, to positive (negative) values, as can be seen in Figure 4.4, where the tangent lines at $(x_0 - h, f(x_0 - h))$, $(x_0, f(x_0))$ and $(x_0 + h, f(x_0 + h))$ ($h > 0$, and very small) are drawn for a minimum at $x = x_0$. If the second derivative vanishes at x_0, however, then the situation is more ambiguous and further examination of the sign of the second derivative at $x_0 \pm h$ is required. A minimum (maximum) obtains in such a situation if the values of $f^{(2)}(x_0 + h)$ and $f^{(2)}(x_0 - h)$ are both positive (negative). If $f^{(2)}(x_0 + h)$ and $f^{(2)}(x_0 + h)$ differ in sign then there is a point of inflexion at x_0.

$\triangleright$
The rate of change in the slope of the tangent (often termed the curvature).

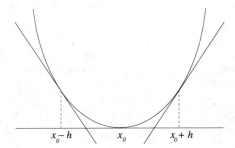

Figure 4.4
The change in the slope of the tangent in the vicinity of a minimum

In summary, a turning point at x_0 corresponds to a

minimum if either $f'(x_0) = 0$ and $f^{(2)}(x_0) > 0$,

or $f'(x_0) = 0$, $f^{(2)}(x_0) = 0$ and $f^{(2)}(x_0 \pm h) > 0$,

maximum if either $f'(x_0) = 0$ and $f^{(2)}(x_0) < 0$,

or $f'(x_0) = 0$, $f^{(2)}(x_0) = 0$ and $f^{(2)}(x_0 \pm h) < 0$,

point of inflexion if either $f'(x_0) = 0$ and $f^{(2)}(x_0) = 0$,

or $f'(x_0) \neq 0$, $f^{(2)}(x_0) = 0$

and $f^{(2)}(x_0 \pm h)$ are of different sign.

Notice that a point of inflexion can occur if $f^{(2)}(x_0) = 0$, even if $f'(x_0) \neq 0$; it is only at a turning point that $f'(x_0) = 0$ is necessary.

Worked example

4.5 Examine the following functions $y = f(x)$ for maxima, minima and points of inflexion in the interval $[-2, 2]$.

(a) $f(x) = x^2$, (b) $f(x) = (x+1)^3$, (c) $f(x) = x^3 - x$.

Solution (a) Since $f'(x) = 2x$ and $f^{(2)}(x) = 2$, the turning point at $(0,0)$ corresponds to a local minimum. At the end-points, the value of $f(x)$ is 4. Hence, maxima occur at $x_0 = \pm 2$, and the local minimum is, in fact, the global minimum.

(b) For this function, $f'(x) = 3(x+1)^2$, $f^{(2)}(x) = 6(x+1)$, and a turning point occurs at $x_0 = -1$. However, the second derivative is zero at this value of x, and since $f^{(2)}(x_0 \pm h) = \pm h$, the point $(-1, 0)$ corresponds to a point of inflexion. The minimum and maximum values of the function occur at the end-points $x = -2$ and $x = 2$, respectively.

(c) Here, $f'(x) = 3x^2 - 1$ and $f^{(2)}(x) = 6x$, and hence turning points occur at $x = \pm 1/\sqrt{3}$. Since $f(1/\sqrt{3}) > 0$ and $f(-1/\sqrt{3}) < 0$, these points correspond to a local maximum and minimum, respectively. The global minimum and maximum occur at $x_0 = -2$ and $x_0 = 2$, respectively. The second derivative is zero at $x_0 = 0$, where $f^{(2)}(x_0 \pm h) = \pm h$; hence this point corresponds to a point of inflexion (notice that $f'(0) \neq 0$).

Problem 4.9

$\triangleright$

See, for example, Alberty and Silbey, Chapter 11.

The $3s$ atomic orbital for the hydrogen-like atom has the following form $\triangleright$:

$$\psi = N\left[27 - 18\left(\frac{Z}{a_0}\right)r + 2\left(\frac{Z}{a_0}\right)^2 r^2\right]e^{-Zr/(3a_0)},$$

where N is a constant. By letting $\sigma = Zr/a_0$, use the chain rule to show that

$$\frac{d}{dr}\psi = \frac{d\psi}{d\sigma}\cdot\frac{d\sigma}{dr} = \frac{NZ}{a_0}\left(-27 + 10\sigma - \frac{2}{3}\sigma^2\right)e^{-\sigma/3}.$$

Hence,

(a) find the maximum and minimum values of ψ for $0 \leq r < \infty$, and identify the values of r associated with the global maximum and minimum values of ψ;

(b) sketch the form of ψ as a function of r, given that $\psi(r) \to 0$ as $r \to \infty$.

Problem 4.10

Using the derivative of the function $h(J)$ given in Problem 4.6, show that the maximum in the fractional occupation number for the rotational energy levels in the vibrational ground state of $^1\text{H}^{35}\text{Cl}$ occurs when $2J + 1 = \sqrt{2T/\Theta_r}$. Given that $\Theta_r = 15.2$ K, determine which J level has the greatest fractional occupation at 298 K.

Problem 4.11

Atkins, p. 426.

The effective potential energy, $V(r)$, for the electron at a distance r from the nucleus in the hydrogen atom ▷ consists of two terms: one arising from the Coulombic attraction between the electron and nucleus; the other from the rotational (angular) motion of the electron. $V(r)$ is given in the form

$$V(r) = -\frac{e^2}{4\pi\epsilon_0 r} + \frac{\ell(\ell+1)h^2}{8\pi^2 m r^2} ,$$

where m and e are the mass and charge of the electron, respectively.

a_0 is defined in Problem 1.6(c).

(a) Show that if $V(r)$ is measured in units of $e^2/(4\pi\epsilon a_0)$ and $r = \rho a_0$, then ▷

$$\mathcal{V}(\rho) = -\frac{\infty}{\rho} + \frac{\ell(\ell+\infty)}{\epsilon\rho^\epsilon} .$$

See Atkins, Figure 13.4.

(b) Assuming $\ell \neq 0$ (the electron occupying an s orbital is excluded), show that there is a turning point at $(\rho_m, \mathcal{V}(\rho_{\updownarrow}))$, where $\rho_m = \ell(\ell+1)$. Identify the nature of the turning point ▷.

Problem 4.12

Alberty and Silbey, p. 416.

The Lennard-Jones potential for describing the nature of the pairwise interaction between neutral atomic or molecular species is taken in the form ▷

$$V(r) = 4\epsilon\left\{\left(\frac{\sigma}{r}\right)^{12} - \left(\frac{\sigma}{r}\right)^6\right\},$$

where the units of ϵ are J.

(a) Determine the turning point, $(r_m, V(r_m))$, of $V(r)$, giving its location, in terms of σ.

(b) Demonstrate by analysing the second derivative of $V(r)$ at r_m that the turning point corresponds to a minimum.

▷

Atkins, Section 12.1.

Problem 4.13

The motion of a particle in a one-dimensional 'box' (line) of length L, and moving under the influence of a constant potential, forms one of the classic quantum mechanical problems for which the corresponding Schrödinger equation can be solved ▷. The wavefunction for this simple system takes the form

$$\psi(x) = \left(\frac{2}{L}\right)^{\frac{1}{2}} \sin\left(\frac{n\pi x}{L}\right).$$

(a) Show that $\psi(x)$ has turning points when $(n\pi x/L)$ is an odd multiple of $\pi/2$: that is,

$$\frac{n\pi x}{L} = (2m+1)\frac{\pi}{2} \quad \text{with} \quad m = 0, 1, \ldots$$

(b) For $n = 3$ give the values of x corresponding to turning points.
(c) Find the second derivative of $\psi(x)$ and identify the nature of each of the turning points in part (b).

The next problem demonstrates the uses of the laws of exponents, the logarithm and the antilogarithm functions all within one problem in chemical kinetics. It is not an easy problem to tackle, and so it is presented in a series of subproblems to make the solution more tractable.

Problem 4.14

The kinetics of the hydrolysis of diethyl butanedioate, A, in the presence of HCl (as a catalyst) can be interpreted in terms of the consecutive first-order reaction A → B → C, in which an ethyl group is lost at each step. As seen in Problem 3.6(c), the concentration of B at time t is given by

$$b = \frac{\alpha \cdot k_1}{k_2 - k_1}\left(e^{-k_1 t} - e^{-k_2 t}\right) \tag{4.11}$$

where α is the initial concentration of A.

(a) Use Equation (4.11) to determine db/dt, and show that db/dt is zero when $k_1 e^{-k_1 t} = k_2 e^{-k_2 t}$.
(b) Take logarithms to the base e of each side of the result in (a), and show that ▷

▷

t_m is the time at which $db/dt = 0$.

$$t_m = \frac{1}{k_2 - k_1} \ln(k_2/k_1) = \ln(k_2/k_1)^{1/(k_2 - k_1)}.$$

(c) Demonstrate that $d^2 b/dt^2 < 0$ at the turning point, where $k_1 e^{-k_1 t_m} = k_2 e^{-k_2 t_m}$.

(d) Show that B achieves a maximum concentration, b_m, given by

$$b_m = \alpha \frac{k_1}{k_2} e^{-k_1} = \alpha \frac{k_1}{k_2} \cdot \frac{1}{e^{k_1 t_m}}.$$

(e) Use the expression for t_m to show that the maximum concentration of B is given by

$$b_m = \alpha \frac{k_1}{k_2} \cdot \frac{1}{(k_2/k_1)^{k_1/(k_2-k_1)}} = \alpha \frac{k_1}{k_2} \cdot \left(\frac{k_1}{k_2}\right)^{k_1/(k_2-k_1)}$$

$$= \alpha (k_1/k_2)^{k_2/(k_2-k_1)}.$$

In many situations, the algebraic method for identifying turning points leads to equations that do not readily admit of a solution in a simple form. In these cases, it may be possible to use computer algebra software (for example, Maple) to carry out the differentiation but then the location of the turning points may have to be carried out numerically. Some aspects of numerical analysis are considered in later chapters but, again, there are always computer algorithms available that may offer a route to the solution required.

4.7   The differentiation of functions of two or more variables: a preview

Before we finish this review of functions of a single variable (through excursions into the fields of power series, complex numbers revisited, and integration), it is appropriate to comment very briefly on the differentiation of functions of two or more variables. The main reason for doing this is that thermodynamics usually comes quite early on in first year chemistry courses, and the need to handle such functions is acute! For example, the entropy function S for a system containing A, B and AB species at a given temperature and pressure depends upon five variables: N_A, N_B, N_{AB}, T and P.

The need for working with functions of this kind arises simply because chemistry takes place in a three-dimensional world subject to the additional constraints of two of the three 'external' variables of temperature, pressure and volume ▷. At the very least, therefore, in the simplest chemical process, we are concerned with two external variables, concentration of a species, and time.

Logically, we should await a more complete development of the calculus of functions of two or more variables before proceeding any further; however, need takes precedence over logic in the traditional teaching of chemistry! This is why we present a short interlude here, in preparation for later developments.

▷

The third external variable is fixed by the equation of state: e.g., $PV = RT$ for 1 mol of an ideal gas.

The partial derivative

As we shall see later, functions of more than two variables cannot be visualized in simple geometrical terms. It is important, therefore, to become used to thinking about functions in an abstract and algebraic way, rather than relying too much on some kind of geometrical view. Derivatives of such functions with respect to one of the (independent) variables are easily found by treating all the other variables as constants and differentiating only with respect to the single variable of interest; such derivatives are termed *partial derivatives*. Thus, if $z = xy + y^2$, then the partial derivative of z with respect to x is found by treating y as a constant. However, to make it clear that several variables are present, we use the notation

$$\left(\frac{\partial z}{\partial x}\right)_y$$

▷
Notice that the differentiation operator now involves the 'curly' form of d.

to remind us that the differentiation is carried out with respect to x, keeping y constant ▷ (the y suffix is often dropped in the contexts where there is no ambiguity in recognizing which variable is held constant).

So in the above example,

$$\left(\frac{\partial z}{\partial x}\right) = y, \quad \text{and} \quad \left(\frac{\partial z}{\partial y}\right) = x + 2y.$$

Problem 4.15

For 1 mol of an ideal gas, the pressure, P is a function of the two variables T and V in the form $P = RT/V$.

(a) Write down $\left(\frac{\partial P}{\partial T}\right)_V$ and $\left(\frac{\partial P}{\partial V}\right)_T$.

(b) Express V as a function of P and T and evaluate $\left(\frac{\partial V}{\partial T}\right)_P$.

(c) Derive an expression for the expansion coefficient, α, of an ideal gas defined by $\alpha = \frac{1}{V}\left(\frac{\partial V}{\partial T}\right)_P$.

▷
Alberty and Silbey, p. 18.

Problem 4.16

The van der Waals equation of state for a gas is of the form ▷

$$P = g(V, T) = \frac{RT}{V - b} - \frac{a}{V^2}.$$

(a) Give the domain of $g(V, T)$.

(b) Determine $\left(\frac{\partial P}{\partial V}\right)_T$, $\left(\frac{\partial P}{\partial T}\right)_V$, and $\left(\frac{\partial^2 P}{\partial V^2}\right)_T$.

(c) Given that the critical point, where the pressure, temperature and volume are P_c, T_c and V_c, respectively, corresponds to a point of inflexion for the function $P = g(V, T)$, show that the equations

$$\left(\frac{\partial P}{\partial V}\right)_{T=T_c, V=V_c, P=P_c} = 0 \text{, and} \left(\frac{\partial^2 P}{\partial V^2}\right)_{T=T_c, V=V_c, P=P_c} = 0$$

lead to the following values of the critical properties ▷:

$$V_c = 3b, \quad T_c = \frac{8a}{27Rb}, \quad P_c = \frac{a}{27b^2}.$$

▷

Remember that, at the critical point, $P_c = RT_c/(V_c - b) - a/V_c^2$.

Summary: This concludes an introduction to the methods for forming first and higher-order derivatives of functions of a single variable. The analysis was then applied to simple mathematical and chemical functions in order to locate maximum and minimum values. A short preview of the differentiation of functions containing two or more variables was also given.

The next chapter is concerned with the topic of differentials which are related to the first derivative of a function; differentials are important in thermodynamics and in error estimation.

5 Differentials – small and not so small changes

Objectives

This chapter shows how

- changes in the independent variable can be used to define approximations to the change in the dependent variable

- such changes may be used in numerical estimation procedures and in the calculation of errors in a chemical context

- extension to two or more variables can be made

Many areas of chemistry – particularly thermodynamics – are concerned with the interrelation between changes in physical properties consequent upon changes in one or more parameters defining the initial state of the system. These changes can be large or small and, for the moment, we shall focus on small changes. The elucidation of the effects of large changes to a chemical system requires the use of integration methods (see Chapter 6), in which the overall change is described in terms of a summation over small changes.

If we have a function $y = f(x)$ which has a derivative $f'(x)$ then, as seen in the last chapter, $f'(x_0)$ is the slope of the straight line which is tangent to $f(x)$ at the specified value of x equal to x_0. The general form of this tangent line is then

$$y = v_0(x) = f'(x_0)x + a$$

where a is a constant (to be determined shortly), and a different tangent line arises for each choice of x_0. The value for the dependent variable is named v_0, with the subscript to indicate the reference point is x_0.

Since the tangent line function touches the function curve at $x = x_0$ then both must have the same value there, and thus we can say that

$$v_0(x_0) = f'(x_0)x_0 + a = f(x_0).$$

The constant a is therefore given by $a = f(x_0) - f'(x_0)x_0$, and the full expression for the tangent line at x_0 is

$$y = v_0(x) = f'(x_0)(x - x_0) + f(x_0). \tag{5.1}$$

The values of $v_0(x)$ and $f(x)$ have both been called y so that the tangent line and the function can be shown on the same graph.

For the case where $y = x^3$, for example (Figure 5.1), the derivative function is $3x^2$ and the equation of the tangent line at $x_0 = 1$ is $y = 3(x - 1) + 1 = 3x - 2$. This tangent line has intercepts on the x and y axes at $x = 2/3$ and $y = -2$, respectively.

It is usual to generalize the expression for $v_0(x)$ given in Equation (5.1). First $(x - x_0)$ is replaced by h, which is regarded as a variable having its origin at the chosen value of x_0; second, as the value selected for x_0 can be any value of x, x_0 is replaced by x, thereby yielding

$$g(x, h) = f'(x)h + f(x) \tag{5.2}$$

as the most general expression for the tangent line. The left-hand side of the equation, previously designated as $v_o(x)$, now becomes $g(x, h)$, which is a function of the two variables x and h. Although a function of this form has not yet been discussed in any detail ▷, its meaning is clear from the context, and can be emphasized by an example.

▷
See Chapter 10.

Worked example

5.1 Find the equation for the tangent line to the function $y = x^3$ and give its form at $x = 1$.

Solution $f'(x) = 3x^2$. Hence from Equation (5.2)

$$g(x, h) = 3x^2h + x^3$$

and so at $x = 1$ the tangent line takes the particular form $g(1, h) = 3h + 1$, and this is illustrated in Figure 5.1.

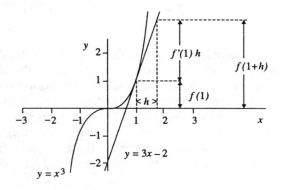

Figure 5.1
The tangent line to the function $y = x^3$ at the point $x = 1$

In terms of the earlier notation, the equation of the tangent line is given by

$$y = 3h + 1 = 3(x - 1) + 1 = 3x - 2,$$

since $h = x - 1$.

5.1   The tangent approximation

Now rearrangement of Equation (5.2) gives as an identity

$$f'(x) = \frac{g(x, h) - f(x)}{h}$$

and, in view of the definition of derivative in Equation (4.1), namely

$$f'(x) = \lim_{h \to 0} \frac{f(x + h) - f(x)}{h},$$

▷
See Chapter 4.

it follows that, for sufficiently small h, the limit of the quotient defining the derivative is given to a good approximation by the quotient itself ▷. Hence,

$$f(x + h) \approx g(x, h) = f'(x)h + f(x). \tag{5.3}$$

For $x = 1$, this equation yields a value for $f(1 + h)$ of $3h + 1$, as indicated in Figure 5.1. This way of estimating the value of a function at a point close to a fixed point is known as the tangent approximation and clearly, unless $f(x)$ does not possess a derivative at the chosen point, we can be quite certain that, for small enough h, the approximation is a good one. Of course it is impossible to give precise rules about the size of h in order to obtain a good approximation for $f(x + h)$. What works depends on the form of the function and the value of x, but an example and some problems will perhaps help to give the feel of what is involved. Before doing this, however, a further shorthand should be noted: the expression $f(x + h) - f(x)$ is the difference in y values, often written as Δy (or Δf), when x is incremented from x to $x + h$ (see Figure 5.2). Thus, the rearranged Equation (5.3) then becomes

$$f(x + h) - f(x) = \Delta y \approx g(x, h) - f(x) = f'(x)h,$$

after substituting for $g(x, h)$ from Equation (5.3). The final term on the right-hand side of this equation corresponds to the change in $f(x)$, when the function is approximated by the tangent line in the vicinity of the point x (see Figure 5.2). $f'(x)h$ is usually written as dy (or df), the *differential* of y. We see that dy, which is easy to calculate from the gradient of the tangent to $f(x)$ at a given point, gives an approximation to the actual change in y values as x is

▷
That is, $\Delta y \approx dy$.

incremented by an amount h ▷. Equation (5.3) shows that the approximate value of the function at the point $x + h$ is $f(x) + dy$ (see also Figure 5.2).

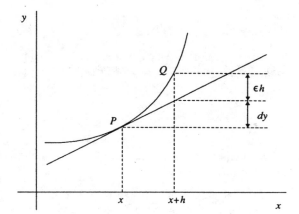

Figure 5.2
The differential, *dy*, and the error, $\epsilon \Delta x$, in its approximation to Δy

Worked example

5.2 Find dy and Δy for the function $y = x^3$, where $x = 4$ and $h = 0.1$.

Solution $f'(x) = 3x^2$. Hence $f'(4) = 48$, and $dy = 48 \times 0.1 = 4.800$, while $\Delta y = f(4.1) - f(4) = 4.921$.

Problem 5.1

Find dy and Δy corresponding to $x = 2, h = 0.5$ and $x = 2, h = 0.05$ for the functions $y = x^2, y = x^3 + x^2 + 1$.

Problem 5.2

Find dy and Δy for the function $y = (6x)^{1/3}$ ▷ when $x = 36$, using first $h = 10$, and then $h = 0.1$.

▷
Use the substitution
$u = 6x$ and the chain
rule for determining
$f'(x)$.

Problem 5.3

The variation of molar heat capacity at constant pressure may be taken in the form, $C_p = a + bT + cT^2$.

(a) Find the derivative of C_p with respect to T, and write down expressions for dC_p and ΔC_p

▷
Alberty and Silbey, p. 52.

(b) for HCl, where appropriate values ▷ of b and c are 1.809×10^{-3} J K^{-2} mol^{-1}, and 15.465×10^{-7} J K^{-3} mol $^{-1}$, respectively, calculate the values of dC_p and ΔC_p, when T changes from 400 K to 410 K.

▷
Alberty and Silbey,
p. 792.

Problem 5.4

The de Broglie wavelength associated with a beam of thermal neutrons at a temperature T can be written as ▷

$$\lambda = \frac{h}{p} = \frac{h}{(mkT)^{\frac{1}{2}}},$$

where $h = 6.626 \times 10^{-34}$ J s is Planck's constant, $m = 1.675 \times 10^{-27}$ kg is the mass of a neutron, and $k = 1.381 \times 10^{-23}$ J K^{-1} is the Boltzmann constant.

(a) Calculate λ at $T = 298$ K and at $T = 288$ K.
(b) Give the expression for $d\lambda$ as a function of T.

▷
Note that ΔT is negative.

(c) Use the expression for $d\lambda$ to estimate the change in wavelength when T is changed from 298 K to 288 K ▷, and compare this estimate with the actual change in wavelength, $\Delta\lambda$.

Problem 5.5

The rate constant of a reaction, k, involving ions in aqueous solution depends upon their charges, and the ionic strength of the solution, I:

$$K = \frac{k}{k^o} = 10^{2Az_Az_B I^{\frac{1}{2}}},$$

where z_A, z_B are the charges of the ions, and A is the Debye–Hückel constant with value 0.509 mol$^{-\frac{1}{2}}$ kg$^{\frac{1}{2}}$, and k^o is the rate constant when activity coefficients are unity ▷.

▷
Atkins, p. 949.

(a) Show that the above equation for K may be written as

$$K = e^{2Az_Az_B I^{\frac{1}{2}} \ln 10},$$

and that $dK = Az_Az_B \ln 10\ I^{-\frac{1}{2}} K\ dI$.

(b) For the base hydrolysis of B = [CoBr(NH$_3$)$_5$]$^{2+}$, A = OH$^-$, where $z_Az_B = -2$, estimate the value of K for I/mol kg$^{-1} = 0.025$ and 0.015, given that for I/mol kg$^{-1} = 0.020$, $K = 0.515$.

5.2   Some further uses of the tangent approximation

The Newton–Raphson method

The tangent approximation may be used to solve non-linear equations (for their real roots at least) and the procedure for doing this is probably best seen by means of an example.

Worked example

5.3   Find a real root of $f(x) = x^5 - 2 = 0$.

Solution   To find a starting point we notice that $f(x) = -1$ at $x = 1$ and $f(x) = 30$ at $x = 2$, so that $f(x)$ must be zero somewhere in between these two values. The root obviously lies closer to $x = 1$ than to $x = 2$, so that a reasonable starting guess might be 1.2. We can then assert using Equation (5.3) that, approximately   at   least,   there   is   a   root   at $1.2 + h$ and $f(1.2 + h) \approx f'(1.2)h + f(1.2) = 0$. A first estimate of h is therefore given by

$$h = -\frac{f(1.2)}{f'(1.2)} \Rightarrow h = -\left(\frac{(1.2)^5 - 2}{5(1.2)^4}\right) = -\frac{0.488\,32}{10.386} = -0.0470.$$

The value of $f(1.2 - 0.047)$, that is $f(1.153)$, is 0.037, which is less than $f(1.2)$, so it seems likely that we are heading in the direction of a solution. We therefore try another iteration starting from $x = 1.153$ to obtain the next estimate for h in the form

$$h = -\frac{f(1.153)}{f'(1.153)} = -0.0013$$

which yields a revised value for the root of $x = 1.149$. The value of $f(1.149)$ is now zero to three decimal places, and this might be a suitable place to stop the iteration $\triangleright$.

$\triangleright$
The number of significant figures is determined in the chemical situation by the precision of the data defining the equation whose root is sought.

The general iterative scheme that we have described is usually called the Newton–Raphson method, and successive estimates of the root are obtained using

$$x_{n+1} = x_n - \frac{f(x_n)}{f'(x_n)} \quad (n = 1, 2, \ldots) \tag{5.4}$$

where x_n is the nth estimate for the solution of $f(x) = 0$. The iteration procedure is stopped when x_{n+1} has converged to x_n to the required accuracy.

Whether or not the sequence of successive x_n values actually converges to a root depends on the starting value chosen for x. However, the algorithm in Equation (5.4) cannot be applied purely mechanically, and care must be taken to check that convergence is going well. In some cases too, the method will fail because there is no real-number solution. Thus an attempt to find the roots of $x^2 + 1 = 0$ would fail because its only roots are $\pm i$ as we shall show in Chapter 8.

Although so far we have used polynomial equations as examples, the method is more generally applicable as we illustrate in the following example.

Worked example

5.4   Find the root of the equation $\cos x - x = 0$.

Solution   Here $f(x) = \cos x - x$, and $f'(x) = -\sin x - 1$. Hence, from the algorithm in Equation (5.4) we obtain

$$x_{n+1} = x_n - \left(\frac{\cos x_n - x_n}{-\sin x_n - 1} \right) = x_n + \frac{\cos x_n - x_n}{\sin x_n + 1}.$$

Since the solution yields the value of x for which $\cos x = x$, we are seeking the point(s) where the curve $y = \cos x$ and the line $y = x$ intersect. A simple sketch of the two functions shows that the point of intersection lies somewhere between $x = 0$ and $x = \pi/2$, so a choice of $x = 1$ is not unreasonable as a starting point, x_1. From this starting point the sequence of estimates $x_2 = 0.7504$, $x_3 = 0.7391$, $x_4 = 0.7391$ is obtained, and we therefore conclude that the root is 0.7931 to four places of decimals.

Problem 5.6

(a) Find the positive root of $2 \sin x = x$.
(b) Explain why the Newton–Raphson method will not converge to a root of $x^2 - 3x + 6 = 0$ starting from $x = 1$.
(c) Apply the Newton–Raphson method to the equation $x^2 - a = 0$ to derive the following algorithm for extracting the square root.

$$x_{n+1} = \frac{1}{2} \left(x_n + \frac{a}{x_n} \right).$$

▷
You will need to work to four decimal places and then round the answer.

Hence find $\sqrt{8}$ to three decimal places, starting from $x_1 = 3$ ▷.
(d) Locate the positions of the nodes in the polynomial part of the hydrogen atom $3s$ atomic orbital function, $(\rho^2 - 6\rho + 6)$, to three decimal places, starting from (i) $\rho = 1$, (ii) $\rho = 4$, (iii) $\rho = 10$. Compare your answers to the values determined in Problem 1.12.

Reformulating the tangent approximation

The tangent approximation is now developed in a manner that provides a preparation for the discussion of power series in Chapter 7.

The first step involves rewriting Equation (5.2) by re-identifying x with x_0 and h with $x - x_0$ to give

$$g(x_0, x - x_0) = f'(x_0)(x - x_0) + f(x_0).$$

▷
Δx was previously designated h.

The increment in x, $(x - x_0)$, is now designated as dx or Δx ▷. It should be noticed, however, that while Δy and dy designate different increments (Figure 5.2), Δx and dx are just two different notations for the same increment. This notation is used when we want to emphasize changes from a chosen value of x.

Hence in the tangent approximation we can re-express Equation (5.3) in the form

$$\Delta y = f(x_0 + \Delta x) - f(x_0)$$
$$= dy + \cdots$$
$$= f'(x_0) \cdot \Delta x + \cdots .$$

It therefore follows that

$$f(x) = f(x_0) + f'(x_0) \cdot (x - x_0) + \cdots , \tag{5.5}$$

where the terms indicated by $\cdots$ are the error terms, often designated by $\epsilon \Delta x$ (see Figure 5.2), that must be be supplied to make the approximate equality of Equation (5.3) exact ▷. Equation (5.5) indicates that, given the values of the function and its first derivative at x_0, the value of the function at a neighbouring point $x = x_0 + \Delta x$ is known to within an error of $\epsilon \Delta x$. Clearly, the smaller the value of Δx, the closer the estimate of the value for $f(x)$ becomes. We shall return to this kind of expansion for functions in a later chapter ▷, when further analysis will provide a formulation of the error term $\epsilon \Delta x$ in terms of the second and higher-order derivatives of $f(x)$.

▷

ϵ, which depends upon Δx, is such that $\lim_{\Delta x - 0} \epsilon = 0$.

▷·

On Maclaurin and Taylor series.

Problem 5.7

For $f(x) = e^x$, take $x_0 = 0$ and use Equation (5.2) to find a two-term approximation for e^x. Compare the value of $e^{0.01}$ obtained in this way with the exact value. Give the percentage error in your calculated result.

Problem 5.8

The volume of a liquid at atmospheric pressure shows significant changes only with temperature: let the relationship be $V = W(T)$.

(a) From the definition show that $dV = W'(T)\Delta T$.
(b) If the volume is $W(T_0)$ at the initial temperature T_0 and $W(T_1)$ at a changed temperature T_1, use Equation (5.5) to show that

$$W(T_1) = W(T_0) + W'(T_0) \cdot (T_1 - T_0) + \cdots$$
$$= W(T_0) + \alpha W(T_0) \cdot (T_1 - T_0) + \cdots ,$$

and identify α, the coefficient of (cubical) expansion.
(c) Use the tangent approximation to demonstrate that $dV = \alpha W(T_0)\Delta T$.
(d) Given that water has a coefficient of cubical expansion of 2.1×10^{-4} K^{-1} calculate dV and the percentage change in volume ▷ when 0.1 dm^3 of water is raised by a temperature of 5 K.

▷

$100 \times dV/V(T_0)$.

Problem 5.9

For a chemical reaction that displays first-order kinetics, the concentration of starting material at time t, a, is given by $a = \alpha e^{-kt}$, where k is the rate constant and α the concentration at $t = 0$. The time, τ, at which $a = \alpha/2$ is called the half-life of the reaction.

Show that

(a) by using Equation (5.5), and taking y as a, x_0 as τ and x as t, a may be approximated by

$$a = \frac{1}{2}\alpha[1 - kt + k\tau];$$

(b) $k\tau = \ln 2$;
(c) the difference between the exact and the approximate values of a at 2τ is $\alpha(\ln 4 - 1)/4$.

5.3   The differential of a function of two variables: a preview

In the case of a function of one variable, $y = f(x)$, the differential dy is given by

$$dy = f'(x) \cdot dx = \frac{dy}{dx} \cdot dx,$$

where now h is (again) replaced by dx.

This idea can be extended to two or more variables – the details of which form the subject matter of Chapter 9. However, a useful preview of the extension of the preliminary ideas of partial differentiation given in Chapter 4 provides a useful working basis (without requiring too much underlying understanding!) for coping with some of the equations of thermodynamics. Thus, for a function of two variables, $z = g(x, y)$, the equivalent of dy for a function of a single variable is dz, which takes the form of a sum of the contributions from each of the two variables:

$$dz = \frac{\partial z}{\partial x}\,dx + \frac{\partial z}{\partial y}\,dy. \tag{5.6}$$

The two partial derivatives are taken with respect to the variable indicated, and each contribution to dz involves the rate of change of z with respect to the chosen variable times the increment in that variable. The formula extends in an obvious way to three or more variables.

Problem 5.10

The volume of a solution of $MgSO_4$ in H_2O depends upon the number of moles of solvent, n_A, the number of moles of solute, n_B, the temperature, T, and the pressure, P. Give an expression for the change in volume, dV, when

(a) all four variables are permitted to vary, taking care to specify the variables that are held constant during each partial differentiation;
(b) T and P are held constant;
(c) T, P, and n_A are held constant.

In all these expressions for dV, the partial derivatives of the volume with respect to the number of moles of solvent or solute are termed *partial molal volumes* – quantities that are important in the application of thermodynamics to the study of solutions.

5.4   Some discussion of the idea of a differential

The idea of a differential as a well-defined quantity has rather fallen out of favour in some areas of mathematics, particularly in those branches which are used in general relativity, and in more advanced kinds of thermodynamics. It would seem prudent, therefore, to indicate where the difficulties lie. No problems arise when dealing with functions of a single variable because in that case it can be shown that the expression $f(x)dx$ is always the differential of some function $F(x)$ such that

$$F'(x) = f(x) \Rightarrow F'(x)dx = f(x)dx \Rightarrow dF(x) = f(x)dx,$$

which is often shortened to $dF = f(x)dx$.

This means that it is possible to give a perfectly precise definition to the quantity $F(x_0 + h) - F(x_0)$ in terms of $f(x)$. As we shall show later this is given in terms of a process called integration which reverses the effect of differentiation. We simply quote the result here and will refer back in due course to explain it in more detail:

$$F(x_0 + h) - F(x_0) = \int_{x_0}^{x_0+h} f(x)dx. \tag{5.7}$$

This form provides a precise definition of ΔF without reference to the tangent approximation as in (5.3). The problems arise, however, in dealing with functions of more than one variable. For ease of exposition we will consider functions of only two variables and in that case if we have a function of the form:

$$a(x,y)dx + b(x,y)dy \tag{5.8}$$

(which is a function of four variables, x, y, dx and dy) there is absolutely no guarantee that there exists a function $z(x,y)$ such that $a(x,y)$ is $\partial z/\partial x$, and so on. Thus although the function of four variables given above is a perfectly well-defined object, it is seldom a differential. Thus it is usually not possible to define Δz in the precise and unambiguous way that ΔF is defined by Equation (5.7). This observation may seem trivially true and it certainly would not matter to practising chemists were it not for the fact that in the historical development of dynamics, and later of thermodynamics, objects like those shown in Equation (5.8) were written as if they were differentials. To distinguish such objects from differentials they were termed improper, inexact or non-total differentials and differentials in consequence were called proper, exact or total differentials. We may regret this confusion but we just have to live with it and remember to guard against it. The confusion here is often compounded by insisting that objects like dx or dy are tiny quantities (often called infinitesimals) but, as we have seen, there is nothing in the mathematics as such that requires them to be any particular size at all. The idea of smallness comes not from the mathematics but from the physics or chemistry of the problem. Thus the pressure–volume work in a process is given by the familiar expression $-pdV$, only if dV is sufficiently small. This is because the pressure must remain the same during the volume change for this expression to define work correctly. In this context too, it is common to write expressions such as

$$\delta W = -pdV \quad \text{or} \quad \text{đ}W = -pdV$$

to indicate that the expression on the right-hand side is not a differential. The problem described here can be rationalized because it turns out to be possible (in all but the most pathological cases) to define a path (or curve) function $t(x,y)$ so that any expression that looks like a differential in x and y can be converted into a differential of the single variable t. Of course the differential is different for each choice of $t(x,y)$ so it is not really possible to speak of *the* differential but once a choice has been made then it is possible to define a change in the function of two variables along a chosen path precisely by means of an integral like that in Equation (5.7) in terms of the single variable t. This last observation, namely that the change is path dependent, corresponds to the assertion in thermodynamics that changes in quantities like work and heat in any process, depend on the path chosen between the initial and final states in the process. Of course there are cases when the form that looks like a differential really is a differential and we shall investigate these cases in more detail later, but in thermodynamics the functions that correspond to differentials are called state functions. Examples are the internal energy function U, the entropy function S and the Gibbs function G. It is worthwhile noting here that while it is always possible to construct a path between two points defining the initial and final states of a system, the physics or chemistry of the problem may forbid certain paths that are mathematically possible. Thus the second law of thermodynamics in this context asserts that certain paths are forbidden when raising the temperature of a system. We shall return to these matters later when we consider partial differentiation in Chapter 10 and multiple integration in Chapter 11.

Summary : The concept of a differential for a function of a single variable has been introduced in order to provide a simple approximation method for estimating changes in the dependent variable when the independent variable is changed by a given amount. When this latter change is small, the differential of the dependent variable can be used for estimating errors and also for finding roots (not necessarily all of them) of polynomials and other kinds of equation. An outline preview of the construction of differentials for functions of more than one variable is given in order to assist in the early assimilation of some thermodynamics.

The next chapter is concerned with the important area of integral calculus, which is developed in terms of an inverse operation to differentiation.

6 Integration – undoing the effects of differentiation

Objectives

This chapter

- develops the idea of integration as an inverse operation to differentiation

- shows how the derivative function can be used to obtain the indefinite integral in simple cases

- demonstrates how some kinds of integral can be simplified through the use of an intermediate variable

- introduces the definite integral by specifying the interval of the integration variable explicitly

- shows how definite integrals with infinite end-points, or discontinuities in the integrand, are evaluated

- demonstrates some simple numerical procedures for evaluating definite integrals

- relates the methodology of integration to explicit chemical situations with examples

In Chapter 4, it was seen that the derivative function $f'(x)$ may be obtained by a limiting process involving the average change of $f(x)$ over a steadily decreasing interval. This kind of procedure is important in defining, for example, the rate of disappearance of a reagent in a chemical reaction. Integration provides us with a tool for reversing the effects of differentiation so that, in a chemical context, given the rate of disappearance of a reagent (the rate of change of concentration with time), we can determine the function describing the variation of the concentration with time.

6.1   The antiderivative function and the $\hat{I}$ operator

▷
See Chapter 4, where
the act of differentiation
is represented by the
symbol $\hat{D}$.

The process of integration can be made more transparent by examining the consequences of differentiating ▷ a third-degree polynomial function to obtain a second-degree polynomial:

$$\hat{D}(ax^3 + bx^2 + cx + d) = (3ax^2 + 2bx + c).$$

Here, it should be noticed that polynomials of degree three, differing only in the value of d, all yield the same polynomial of degree two under the act of differentiation. This observation of a many-to-one correspondence between third- and second-degree polynomials under differentiation is easily generalizable, since the differentiation of functions of the form $F(x) + C$, differing only by the constant C, all yield the same function $f(x)$:

$$\hat{D}(F(x) + C) = \frac{d}{dx}F(x) \equiv F'(x) = f(x). \tag{6.1}$$

The constant C plays the same role as d in the third-degree polynomial functions described above.

▷
The function differ only in
the value of C.

The process of reversing the procedure given in Equation (6.1) is described as integration, and is effected by the operator $\hat{I}$, such that $\hat{I}f(x) = F(x) + C$ ▷. To re-iterate: the reversal of the many-to-one process of differentiation yields a *set* of *antiderivative* functions $F(x) + C$, since different values of C correspond to different antiderivative functions. The *indefinite integral* of $f(x)$, $\hat{I}f(x)$, is usually expressed in the form

$$\hat{I}f(x) \equiv \int f(x)\, dx = \int F'(x)\, dx = F(x) + C,$$

where the derivative of $F(x) + C$ must be equal to $f(x)$, as given in Equation (6.1): a result that should be checked after every integration!

It is also important to notice that, because of the link between $F'(x)$ and $f(x)$, the differential of $F(x)$, given by $dF = F'(x)dx$, is equal to $f(x)dx$. However, although the differential of the antiderivative function is always defined for functions of a single variable, this is rarely the case for functions of two or more variables ▷.

▷
As seen in Chapter 5,
and as we shall see in
Chapter 10.

Returning now to the problem of determining antiderivative functions, suppose, for example, that integration of the function $f(x) = 2x$ is required. A possible antiderivative function is $F(x) = x^2$, since differentiation of x^2 yields $2x$. However, the set of *all* functions differing from x^2 by a constant forms the indefinite integral of $f(x)$: $\{x^2 + 1,\ x^2 + 0.1,\ x^2 - 3.7,\ \ldots\}$. This set of functions may be written as $\{x^2 + C : C \in \mathbb{R}\}$ or, more symbolically, as $x^2 + C$. Plots of selected antiderivative functions are given in Figure 6.1.

Thus, lists of integrals of simple functions are constructed by noting that the integral of the derivative function is just the original function plus a constant of integration, C. In Table 6.1, a function in the first column is obtained by differentiating the corresponding function in the second column.

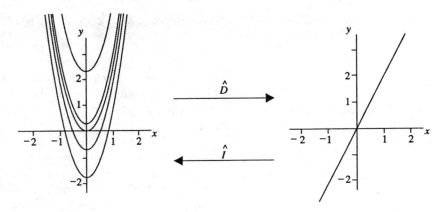

Figure 6.1
Plots of $y = f(x)$ and
$y = F(x) + C$ for $f(x) = 2x$
and $F(x) = x^2$

▷
A more extensive
tabulation is given in
Appendix 3.

Thus the integral of a function in the first column is simply found by adding C to the corresponding function in the second column ▷.

Table 6.1

$f(x)$	— integrate →	$F(x)$
$2x$		x^2
$\sin x$		$-\cos x$
$\cos x$		$\sin x$
e^x		e^x
$1/x$		$\ln x$
$x^n\ (n \neq -1)$		$x^{n+1}/(n+1)$
$\sec^2 x$		$\tan x$
$f(x)$	← differentiate —	$F(x)$

This procedure of building up lists of integrals is of limited use in practice, because the integral of the function required usually emerges from the analysis, and it is not immediately recognized as a derivative function. This is why the skill necessary to determine integrals of functions can only be built up by experience and practice. Some of the strategies required are described in the sections which follow but, in this introductory text, it is impossible to be complete. We shall survey some of the procedures that are usefully deployed within a chemical context, most of which involve the conversion of an integral into a form which can be recognized as being related to one of the standard forms. There are, of course, some integrals which cannot be evaluated in closed form and, although we know a form for $F(x)$ exists in principle (unless $f(x)$ is very odd, which is unlikely in a chemical situation!), we can always approximate it numerically in some interval of x, using a procedure considered later in this chapter. The problem is that, if an expression for $F(x)$ cannot be found, we do not know if an answer either exists in closed form, or whether we are simply unable to find it!

There are also other integrals which require more powerful techniques for their evaluation. For example, the integration of a function of a complex variable, which forms a whole field of mathematics lying outside the scope of this introductory text, provides a technique for evaluating certain kinds of integral that arise in scattering problems (electron–molecule, photon–molecule, etc.). For the present, however, we are concerned with more easily assimilable procedures for integral evaluation.

Worked example

6.1 Show that $\dfrac{d}{dx}\ln(1+x) = \dfrac{1}{1+x}$, and hence deduce that

$$\int \frac{1}{1+x}\,dx = \ln(1+x) + C = \ln\left[A(1+x)\right], \text{ where } C = \ln A.$$

Solution Consider the function $y = \ln(1+x)$, and let $u = 1+x$. Thus, using the chain rule for differentiation

$$\frac{dy}{dx} = \frac{dy}{du}\cdot\frac{du}{dx} = \frac{1}{u}\cdot 1 = \frac{1}{1+x}\,.$$

Hence, we can identify $\ln(1+x)$ with $F(x)$ and $1/(1+x)$ with $f(x)$, so that

$$\int \frac{1}{1+x}\,dx = \ln(1+x) + C = \ln[A(1+x)]\,.$$

Problem 6.1

Show that

(a) $\dfrac{d}{dx}e^{x^2} = 2xe^{x^2}$, hence find $\displaystyle\int xe^{x^2}\,dx$

(b) $\dfrac{d}{dx}\ln(\ln x) = \dfrac{1}{x\ln x}$, hence find $\displaystyle\int \frac{1}{x\ln x^2}\,dx$

(c) $\dfrac{d}{dx}x^{\frac{1}{2}} = \dfrac{1}{2}x^{-\frac{1}{2}}$, hence find $\displaystyle\int \frac{1}{x^{\frac{1}{2}}}\,dx$.

As noted above, and seen through the further examples considered in Problem 6.1, the procedure of forming lists of functions and their derivatives can always be used to augment the earlier table of indefinite integrals. In general, however, we are usually faced with the problem of integrating some function $f(x)$ ▷:

▷
$f(x)$ is termed the integrand.

$$\int f(x)\,dx$$

where the antiderivative function $F(x)$ is not at all obvious to specify. It is necessary, therefore, to spend a little time exploring the various techniques that can be deployed in determining the integral of a given function.

Further properties of the $\hat{I}$ operator

In discussing the operation of integration in terms of antidifferentiation, we can make use of an important property of the $\hat{D}$ operator. If $\hat{D}$ is applied to a function, $H(x)$, that can be written as a linear combination of functions then, from the properties of the derivative stated in Chapter 4, we have

$$\hat{D}H(x) = \hat{D}\{F(x) + G(x)\} = \hat{D}F(x) + \hat{D}G(x) = B(x), \tag{6.2}$$

where $B(x) = f(x) + g(x)$. But we know there are functions $f(x)$ and $g(x)$ such that

$$\hat{D}F(x) = f(x), \quad \hat{D}G(x) = g(x),$$

and

$$\hat{I}f(x) = F(x), \quad \hat{I}g(x) = G(x).$$

Thus, from Equation (6.2), the antiderivative of $B(x)$ is
$H(x) = F(x) + G(x) = \hat{I}f(x) + \hat{I}g(x) = \hat{I}\{f(x) + g(x)\}$.

$\triangleright$
The square root operator is not linear, in that clearly $\sqrt{4+16} = \sqrt{20} \neq 2+4$.
$\triangleright$
The solutions of which determine the physical properties associated with electronic and nuclear motions.

It therefore follows that $\hat{I}$ has the same property as $\hat{D}$ when applied to a sum of functions, in that the operator has the same effect as when it is applied to each function in turn and then the results are summed. We say that both operators are *linear* operators $\triangleright$. Operators of this form are of great importance in quantum mechanics and in the theory of group representations – the former through the construction of the (linear) Hamiltonian operator and thence the Schrödinger equation $\triangleright$; the latter through the representation of symmetry operations, on a crystalline unit cell or the framework of a molecule, in which the positions of equivalent nuclei are permuted. We return to the discussion of symmetry theory after having reviewed the relevant aspects of the theory of matrices in Chapter 13. In the meantime, we return to a discussion of the different methods used for evaluating integrals.

6.2   Methods for evaluating integrals

First, the simplest method of transforming $f(x)$ using algebraic or trigonometric identities so that the resulting integral can be recognized either directly, or be subjected to one or more of the other techniques; second, the substitution method, involving the introduction of an intermediate variable as a first step in transforming the given integral into a form which may be more amenable to finding the antiderivative function; third, the use of partial fractions when $f(x)$ is of the form of a rational polynomial function; and fourth, integration by parts – the application of which is usually easy to recognize. Irrespective of the method used, it is wise to check that the derivative of $F(x) + C$ really does yield the function $f(x)$ being integrated.

If all of the usual methods fail to achieve a tractable integral, then a numerical approach may have to be used.

Rearrangement of the integrand

A good example of this kind of technique is provided by $\int \cos^2 x \, dx$. Here the identities $\cos^2 x + \sin^2 x = 1$ and $\cos^2 x - \sin^2 x = \cos 2x$ are combined to yield the expression $\cos^2 x = \frac{1}{2}(1 + \cos 2x)$, thus enabling the integral to be written as a sum of two integrals – one of which can be performed immediately as it is in standard form:

$$\frac{1}{2}\int dx + \frac{1}{2}\int \cos 2x \, dx = \frac{x}{2} + \frac{1}{2}\int \cos 2x \, dx + C.$$

This leaves a simpler integral to determine than the original one and, in fact, it is now easy to guess that the antiderivative function is $\frac{1}{2}\sin 2x$, as can be seen by differentiation $\triangleright$, so it follows that

$$\int \cos^2 x \, dx = \left(\frac{x}{2} + \frac{1}{4}\sin 2x\right) + C = F(x) + C.$$

Differentiation of $F(x)$ yields

$$F'(x) = \frac{1}{2} + \frac{1}{2}\cos 2x = \cos^2 x = f(x),$$

which confirms that the integration has been carried out correctly.

$\triangleright$
In general, the antiderivative function associated with cos px is $(1/p)$ sin px.

Problem 6.2

(a) Use the identity for $\cos 3x$ given in Problem 2.11 to rewrite $\int \cos^3 x \, dx$ in terms of integrals involving cos px and hence, using the marginal hint above, find the antiderivative function associated with $\cos^3 x$.

(b) Use the identity $\cos^2 x = \frac{1}{2}(1 + \cos 2x)$ to express $\int \cos^4 x \, dx$ in terms of integrals over cos px, and hence determine the integral of $\cos^4 x$.

6.3   The substitution method

In trying to find a function $F(x)$ such that $F'(x) = f(x)$, it is sometimes advantageous to change the name of the integration variable by a judicious form of substitution, in order to produce an integral in simpler form. Consider, for example, the integral

$$I = \int f(x)\,dx = \int \frac{x^2}{(1-x)^{1/2}}\,dx, \tag{6.3}$$

in which the denominator of $f(x)$ needs reformulating before the antiderivative function can be recognized. This can be achieved by at least the three substitutions $u = g(x) = (1-x)^{\frac{1}{2}}$, $u = g(x) = 1-x$, or $x = k(u) = \sin^2 u$. However, before investigating any of these substitutions in detail, we need to consider the general principles underlying any change of variable.

There are two basic strategies for dealing with the integration of a given function, $f(x)$, using the substitution method. For the first two substitutions itemized above, $f(x)$ can be written in the form of a function of a function. In this form, part of $f(x)$ is identified as $u = g(x)$, thereby permitting the integrand to be reformulated in terms of a function of the new integration variable u, since $f(x) = h(g(x)) = h(u)$. For the third form of substitution, a straight change of variable can be made by setting $x = k(u)$, so that the function of u is written in the form $f(k(u)) = w(u)$. In either of these situations, it remains to deal with the change in the instructions relating to the integration variable, designated by the differential dx under the integral sign. As seen in Chapter 5, if u is expressed in terms of x through the relation $u = g(x)$, then

$$du = \frac{du}{dx}\,dx,$$

▷
Remember that since
$u = g(x)$,
$du/dx = g'(x)$.

thus enabling dx to be expressed in terms of du since du/dx ▷ is usually readily expressed in terms of u. Alternatively, if x is expressed in terms of u by means of the function $x = k(u)$, then

$$dx = \frac{dx}{du}\,du,$$

and no inversion of the equation is necessary to express the differential dx in terms of the new variable.

We now return to the evaluation of the integral in Equation (6.3) using the three different choices for the substitution given above.

Worked example

6.2   Evaluate $I = \int \dfrac{x^2}{(1-x)^{1/2}}\,dx$.

Solution (a)   A possible substitution is given by $u = g(x) = (1-x)^{1/2}$. The x^2 term in the integrand is then expressed in terms of u by solving the equation $u = g(x)$ for x in terms of u:

$$u^2 = 1 - x \Rightarrow x = 1 - u^2.$$

Furthermore, since

$$\frac{du}{dx} = \frac{1}{2} \cdot -1(1-x)^{-1/2} = -\frac{1}{2u}$$

$$\Rightarrow du = -\frac{1}{2u} \, dx \Rightarrow dx = -2u \, du \, .$$

Thus, substituting for the variable u reduces the the original integral to a sum of simple integrals involving powers of the new variable

$$I = \int \frac{x^2}{(1-x)^{1/2}} \, dx = \int \frac{(1-u^2)^2}{u} \cdot (-2u) \, du$$

$$= -2 \int (1 - 2u^2 + u^4) \, du = -2 \int du + 4 \int u^2 \, du - 2 \int u^4 \, du \, .$$

Each of these integrals yields an integration constant, and the sum of the three constants is designated by C:

$$I = -2\left(u - \frac{2}{3}u^3 + \frac{u^5}{5}\right) + C = -2u\left(1 - \frac{2}{3}u^2 + \frac{u^4}{5}\right) + C \, . \qquad (6.4)$$

Thus, on reverting to the original variable, x, I may be written in the form

$$I = -2(1-x)^{1/2}\left(1 - \frac{2}{3}(1-x) + \frac{1}{5}(1-x)^2\right) + C \, .$$

Solution (b) The original substitution was made in (a) in order to eliminate the square root from the integrand (often a good strategy); however, the simpler substitution of the first kind $u = 1 - x$ yields a different route through to the same answer, except that integrals over u involve fractional powers of u:

$$I = \int f(x) \, dx = \int \frac{(1-u)^2}{u^{\frac{1}{2}}} \cdot -du \, ,$$

since $du = (du/dx)dx = -dx$. Thus,

$$I = -\int \frac{(1 - 2u + u^2)}{u^{\frac{1}{2}}} \, du = -\int (u^{-\frac{1}{2}} - 2u^{\frac{1}{2}} + u^{\frac{3}{2}}) \, du$$

$$\Rightarrow I = -2\,u^{\frac{1}{2}} + \frac{4}{3}u^{\frac{3}{2}} - \frac{2}{5}u^{\frac{5}{2}} + C \, ;$$

that is,

$$I = -2u^{\frac{1}{2}}\left(1 - \frac{2}{3}u + \frac{1}{5}u^2\right) + C$$

$$\Rightarrow I = -2(1-x)^{\frac{1}{2}}\left(1 - \frac{2}{3}(1-x) + \frac{1}{5}(1-x)^2\right) + C \, ,$$

on substituting for x.

Solution (c)

▷

A trigonometrical substitution suggests itself here as a way of eliminating the awkward part of $f(x)$ that contains the square root.

It is also possible for the present example to make a substitution of the second kind, in which x is substituted for a function of u ▷. Suppose we take $x = \sin^2 u$, with $dx = 2 \sin u \cos u \, du$, then $(1 - x)^{\frac{1}{2}} = (1 - \sin^2 u)^{\frac{1}{2}} = \cos u$ and $x^2 = \sin^4 u$: thus,

$$I = \int \frac{\sin^4 u}{\cos u} \cdot 2 \sin u \cos u \, du = 2 \int \sin^5 u \, du .$$

This integral looks more menacing than the original integral posed in terms of the variable x! However, integrals involving *odd* powers of the sine function are, in fact, easy to evaluate after repeated use of the identity $\sin^2 u = 1 - \cos^2 u$, so as to leave $\sin u$ times a power of $(1 - \cos^2 u)$:

$$I = 2 \int (1 - \cos^2 u)^2 \sin u \, du .$$

This integral is evaluated using the substitution $w = \cos u$, so that $dw = - \sin u \, du$:

$$I = -2 \int dw + 4 \int w^2 \, dw - 2 \int w^4 \, dw = -2w + \frac{4}{3} w^3 - \frac{2}{5} w^5 + C$$

$$= -2 \cos u + \frac{4}{3} \cos^3 u - \frac{2}{5} \cos^5 u + C$$

$$= -2 \cos u \left(1 - \frac{2}{3} \cos^2 u + \frac{1}{5} \cos^4 u \right) + C .$$

▷

If u is expressed in terms of x using the substitution $x = \sin^2 u$, then a polynomial in $\cos(\arcsin x^{\frac{1}{2}})$ is obtained; however, it is clear that the latter function is just a disguised form of $(1 - x)^{\frac{1}{2}}$.

Thus, returning to the original variable x, using $\cos u = (1 - \sin^2 u)^{\frac{1}{2}} = (1 - x)^{\frac{1}{2}}$ ▷, yields the expected result

$$I = -2(1 - x)^{\frac{1}{2}} \left(1 - \frac{2}{3}(1 - x) + \frac{1}{5}(1 - x)^2 \right) + C$$

which is the same result as before.

The three substitutions considered in the above example show that the amount of work required in evaluating the indefinite integral depends to a large extent on the choice of substitution, and this is why it is a good idea to gain practice in using the substitution method in its various forms. Not all substitutions end up as being sensible: for example, the substitution $x = \cosh u$ leads to an integral which is not too difficult to evaluate, but the resulting function, $F(x)$, does not have a friendly appearance – although, of course, it must be the same function that we had before to within a constant. Another point worth noting here is that one substitution may lead to a simpler looking integral which may require one or more steps of further substitution before the integral is in the recognizable form of a standard integral.

In summary, using the substitution method to change an integral into a more tractable form necessitates a certain amount of experience in making the decision as to whether an algebraic or trigonometric substitution is the more appropriate. Certain kinds of integral, however, may require the use of one or more of the techniques discussed in later sections.

> Problem 6.3
>
> For each of the integrals below,
> 1. Give the range of x values for which the integrand, $f(x)$, is defined.
> 2. Use the substitutions suggested and, for practice, try other substitutions that suggest themselves.
>
> (a) $\displaystyle\int \frac{x}{(1 - x^2)^{1/2}} \, dx$, using $x = \sin\theta$
>
> (b) $\displaystyle\int x(x + 4)^{1/2} \, dx$, using first $(x + 4) = u$ and then $(x + 4)^{\frac{1}{2}} = w$
>
> (c) $\displaystyle\int \frac{1}{x - x^{1/2}} \, dx$ using $x^{\frac{1}{2}} = u$,
>
> (d) $\displaystyle\int \frac{dx}{a^2 + x^2}$ using $x = a\tan\theta$
>
> (e) $\displaystyle\int \frac{dx}{(a^2 + x^2)^{\frac{5}{2}}}$ using first $x = a\tan\theta$ and then the result ▷ from
>
> Problem 6.2(a).
>
> (f) $\displaystyle\int \frac{1}{\cosh x} \, dx = \int \operatorname{sech} x \, dx$ using $u = e^x$ and then $u = \tan\theta$ ▷
>
> (g) $\displaystyle\int \frac{e^x}{e^x - 1} \, dx$ using both $e^x - 1 = u$ and also $e^x = u$,
>
> (h) $\displaystyle\int \cos^3 x \, \sin^3 x \, dx$ using $u = \sin x$.

▷
See also Equation
(A2.23) in Appendix 2.

▷
$\cosh x = (e^x + e^{-x})/2$.

A useful result

There is one useful application of the substitution method that arises in the integration of a quotient of functions, when the derivative of the denominator is equal to a constant times the numerator. Suppose, for example, that we require to evaluate the integral

$$I = \int \frac{g(x)}{f(x)} \, dx \,,$$

where $g(x) = kf'(x)$, and k is a constant. Use of the substitution $u = f(x)$ transforms the integral into the standard form

$$I = \int k\frac{du}{u} = k \ln u + C \,, \tag{6.5}$$

which is easy to evaluate, since $du = f'(x)\,dx$. Thus

$$I = \int \frac{g(x)}{f(x)}\,dx = \int kf'(x)/f(x)\,dx = k\ln(f(x)) + C.\qquad(6.6)$$

The steps involved in Equation (6.5) illustrate an important property in carrying out an integration process, in that a constant factor in the integrand may be taken outside the integral sign ▷ since in the last step ln u may be written as the integral of $1/u$. Hence,

$$I = \int k\frac{du}{u} = k\ln u + C = k\int \frac{du}{u} + C.$$

▷

So, in general,
$\int kf(x)\,dx = k\int f(x)\,dx$.

Now the steps outlined in the above integration scheme necessitate that $f(x) > 0$, in order that the logarithm function is defined. If $f(x) < 0$ then it is easy to use the preliminary substitution $g(x) = -f'(x)$, which leads to the expression $I = \ln(-f(x)) + C$ for the integral ▷. Thus, both cases can be accommodated by writing $I = \ln(|f(x)|) + C$, where $|f(x)|$ is a slight modification of the modulus function introduced in Chapter 2 (the value of x such that $f(x) = 0$ is not permitted here, as the integrand then becomes infinite or undefined).

▷

$-f(x) > 0$.

▷

Remember the definition of tan x.

> **Problem 6.4**
>
> Identify $f(x)$ and $f'(x)$ in each of the following integrands, and hence evaluate the following integrals, taking constant factors outside the integral sign where appropriate ▷:
>
> $$\int \frac{x}{x^2+4}\,dx, \quad \int \frac{1}{x+1}\,dx, \quad \int \tan x\,dx, \quad \int \frac{1}{x\ln x}\,dx.$$

6.4   Integrals involving rational polynomial functions

Use of partial fractions

In several applications of the calculus to chemistry (for example, kinetics) it is necessary to integrate a rational polynomial function, which is in the form of a quotient of two polynomial functions:

$$\int \frac{P(x)}{Q(x)}\,dx.$$

If the degree of $P(x)$ is less than that of $Q(x)$, and $Q(x)$ factorizes, then the simplest route to evaluating the integral involves the use of *partial* fractions to rewrite the integral as a sum of simpler integrals. For example, if $P(x) = 7x + 6$, and $Q(x) = x^2 + 3x + 2 = (x + 2)(x + 1)$, then $P(x)/Q(x)$ is expressed in the form

$$\frac{A}{(x + 2)} + \frac{B}{(x + 1)},$$

where A and B have to be determined. Thus, on rewriting the right-hand side of this expression in terms of a common denominator,

$$\frac{7x + 6}{x^2 + 3x + 2} = \frac{A}{(x + 2)} + \frac{B}{(x + 1)} = \frac{A(x + 1) + B(x + 2)}{(x + 2)(x + 1)}$$
$$= \frac{x(A + B) + A + 2B}{x^2 + 3x + 2}.$$

Comparison of the coefficients of x and the constant terms in the numerators of the first and last expressions in the above equation then yields $A + B = 7$ and $A + 2B = 6$, from which it is seen that $A = 8$ and $B = -1$. Hence,

$$\int \frac{7x + 6}{x^2 + 3x + 2} \, dx = 8 \int \frac{1}{x + 2} \, dx - \int \frac{1}{x + 1} \, dx. \tag{6.7}$$

Both integrals on the right-hand side can be evaluated using the substitutions $u = x + 2$ and $u = x + 1$, respectively, thus enabling the original integral to be written as

$$\int \frac{7x + 6}{x^2 + 3x + 2} \, dx = 8 \ln(x + 2) - \ln(x + 1) + C = \ln D \frac{(x + 2)^8}{(x + 1)},$$

where $C = \ln D$.

Problem 6.5

Use partial fractions to evaluate

(a) $\int \dfrac{dx}{(a - x)(b - x)}$, where a and b are constants, and

(b) $\int \dfrac{x}{(x + 1)(x - 2)} \, dx$.

The integration of other kinds of rational function, like

$$\frac{x}{(x^2 + 2)(x + 1)^2},$$

where the degree of $Q(x)$ is still larger than that of $P(x)$, but $P(x)$ contains either quadratic factors or repeated linear factors, needs a slightly different treatment. When attempts are made to put such expressions into partial frac-

▷

Salas and Hille, Section
8.2.

tion form, the general principles are the same, but the rules are different ▷. In
the present example

$$\frac{P(x)}{Q(x)} = \frac{Ax + B}{x^2 + 2} + \frac{C}{(x+1)^2} + \frac{D}{x+1}$$

using the general rules given in Salas and Hille. The formation of a common
denominator for the right-hand side of the above equation, followed by a
comparison of the powers of x^3, x^2, x^1, x^0 on both sides of the equality, yields
four equations which may be solved for A, B, C, D.

Problem 6.6

(a) Write down the four equations determining the coefficients
$A, B, C,$ and D in the partial fraction expansion of

$$\frac{x}{(x^2 + 2)(x + 1)^2}$$

and verify that $A = -\frac{1}{9}$, $B = \frac{4}{9}$, $C = -\frac{3}{9}$, $D = \frac{1}{9}$.

(b) Evaluate

$$\int \frac{x}{(x^2 + 2)(x + 1)^2} \, dx,$$

by writing the integral as

$$\int \frac{Ax + B}{x^2 + 2} \, dx + \int \frac{C}{(x + 1)^2} \, dx + \int \frac{D}{x + 1} \, dx,$$

and using Equation (6.6) and the result of Problem 6.3(d) for the
first integral, and the substitution method for the remaining
integrals.

If the degree of $P(x)$ is greater than or equal to that of $Q(x)$, then the first
step involves carrying out a division process which leads to two or more
integrals to evaluate, as seen in the following example:

$$\int \frac{x^2}{x^2 + 1} \, dx = \int \frac{(x^2 + 1 - 1)}{x^2 + 1} \, dx = \int \left\{ 1 - \frac{1}{x^2 + 1} \right\} dx$$

$$= x - \int \frac{1}{x^2 + 1} \, dx = x - \arctan x + C,$$

where the integral in the penultimate step is available from Problem 6.3(d),
taking $a = 1$.

It should be noted that the division process (where necessary) enables
$P(x)/Q(x)$ to be written as a sum of a polynomial function and another
rational function $R(x)/Q(x)$, where the degree of $R(x)$ is less than that of $Q(x)$.

Problem 6.7

For $P(x) = x^3$ and $Q(x) = x^2 + 3x + 2$

(a) show that

$$\frac{P(x)}{Q(x)} = x - \frac{3x^2 + 2x}{x^2 + 3x + 2}$$

after one step of division;

(b) repeat the division process for the second rational function in (a) above, and show that

$$\int \frac{P(x)}{Q(x)} \, dx = \frac{x^2}{2} - 3x + \int \frac{7x + 6}{x^2 + 3x + 2} \, dx + C;$$

(c) use Equation (6.7) to eliminate the integral on the right-hand side of (b), and thus give an expression for $\int \frac{P(x)}{Q(x)} \, dx$;

(d) check your result by differentiation.

6.5   Integration by parts

Integrals for which the integrand is a product of functions are sometimes amenable to *integration by parts*. In this method, the initial integral is replaced by another integral which may be more tractable; in some cases, several applications of integration by parts (and even substitutions) may be necessary before the antiderivative function can be recognized. The principle of the method follows directly from the formula for the derivative of a product of two functions

$$\frac{d}{dx} \{ f(x) \, g(x) \} = f'(x) \, g(x) + f(x) \, g'(x),$$

as stated in Chapter 4. Thus integration of this equation yields the expression

$$f(x) \, g(x) = \int f'(x) \, g(x) \, dx + \int f(x) \, g'(x) \, dx \Rightarrow \int f(x) \, g'(x) \, dx$$

$$= f(x) \, g(x) - \int f'(x) \, g(x) \, dx,$$

which in differential notation may be written as

$$\int u \, dv = u \, v - \int v \, du,$$

where $u = f(x)$, $v = g(x)$, $du = f'(x) \, dx$, $dv = g'(x) \, dx$.

At first sight it is not at all clear how this obscure manipulation of an integral can be of any help; however, with a judicious choice of functions, the above identity can be used to produce a simpler integral to evaluate. In many instances it is quite obvious which part of the integrand to ascribe to $g'(x)$ – simply because this function is easier to integrate than $f(x)$. An example and some problems in using the method are now given.

Worked example

6.3   Evaluate

$$\int x^2 e^{-x} dx.$$

Solution   In this example the two obvious candidates for $g'(x)$ are x^2 or e^{-x}, as both can be easily integrated; however, if the former choice is taken then a more complicated integral emerges by applying the formula (one in which x^3 now appears alongside e^{-x} in the integrand). Let us take the latter choice, where $g'(x) = e^{-x} \Rightarrow g(x) = -e^{-x}$; $f(x) = x^2 \Rightarrow f'(x) = 2x$. Hence on using the formula, the original integral is replaced by the expression

$$-x^2 e^{-x} + 2 \int x e^{-x} dx$$

which, as anticipated, contains a simpler-looking integral to evaluate.

Problem 6.8

Identify appropriate choices for $f(x)$ and $g'(x)$, and use the method of integration by parts to evaluate the following integrals:

$$\int x e^{-x} dx, \quad \int x^2 e^{-x} dx, \quad \int x \sin x\, dx, \quad \int x^2 \ln x\, dx,$$

$$\int \sin x \cos x\, dx, \quad \int \ln x\, dx.$$

Hint: In the last integral take $g'(x) = 1$.

Sometimes there are integrals that look as though they might yield to the method of integration by parts. The integral

$$\int x e^{x^2} dx$$

is not amenable to this approach, whereas the method of substitution, based on $u = x^2$, provides the simplest method for its evaluation.

6.6   The definite integral

▷
Salas and Hille,
Section 5.4.

The definite integral is one in which the integration is restricted to those values of x lying in a specific interval, say $a \leq x \leq b$. The result is a number that depends only upon a and b, and not x. An important theorem of calculus ▷ enables us to prove that the value of the definite integral

$$I = \int_a^b f(x) \, dx$$

is just the difference in the values of the antiderivative function $F(x)$ evaluated at the end-points of the interval $[a, b]$; that is

$$I = \int_a^b f(x) \, dx = F(b) - F(a), \tag{6.8}$$

which is generally written as $[F(x) + C]_a^b$. It is important to notice that the definite integral obtained from the indefinite integral of $f(x)$, does not depend upon the integration constant, C, and if $f(x) \geq 0$ then I represents the area under the curve $y = f(x)$ between $x = a$ and $x = b$ as indicated in Figure 6.2.

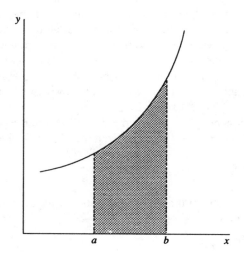

Figure 6.2
Geometrical representation of $\int_a^b f(x) \, dx$, where $f(x) \geq 0$ in the interval $[a, b]$

If, on the other hand, $f(x) \leq 0$ in the interval, then the definite integral is the negative value of the area that lies below the x-axis. For some functions, where $f(x)$ has positive and negative values in $[a, b]$, the definite integral may have a positive, zero or negative value. For example, the definite integral of $\cos x$ over the interval $[0, 2\pi]$ is zero, because of the cancellation of the positive and negative contributions.

> **Problem 6.9**
>
> Evaluate $\displaystyle\int_0^{\frac{\pi}{2}} \cos x \, dx.$

The definition of the definite integral may seem a little strange; however, if the lower limit, a, of the interval is fixed, and the upper limit, x, is allowed to vary, then the resulting value for the definite integral

$$\int_a^x f(t) \, dt = F(x) - F(a) = F(x) + C,$$

▷

The name of the integration variable is of no conseqence in determining the value of the integral.

depends upon the choice for x ▷. The constant value $-F(a)$ has been written as C.

It is clear now that $F(x) + C$ is just the indefinite integral of $f(x)$, with the property that $F'(x) = f(x)$. This means that

$$\hat{D}\int_a^x f(t) \, dt = \hat{D}(F(x) + C) = f(x).$$

This equation displays the important result that the indefinite integral can be written formally as a definite integral, but with a variable upper limit – thus confirming that the indefinite integral is a function, as expected.

As far as definite integrals are concerned, they can be evaluated either by finding the difference in values of the antiderivative function at $x = b$ and $x = a$, as given in Equation (6.8), or by using a numerical approach by finding the net (signed) area under the curve $y = f(x)$ in the interval $a \le x \le b$ ▷. The estimate for this area can be found to arbitrary accuracy by numerical means, and the fact that $F(x)$ is unknown is not important. The numerical approach (discussed in a later section) is especially important in those cases where it is not possible to write down a function $F(x)$ in closed form in order to evaluate $F(b) - F(a)$. For example, the *error function*

▷

If $f(x)$ is negative in one or more subintervals, then the corresponding contributions to the definite integral will carry a negative sign.

$$\text{erf}(x) = \frac{2}{\sqrt{\pi}} \int_0^x e^{-t^2} \, dt,$$

which appears in the study of statistics (Chapter 12), is in this category, as is the integral

$$\int_0^t \frac{x^4 e^x}{(e^x - 1)^2} \, dx$$

▷

Alberty and Silbey, p. 586.

that arises in working through the analysis of the Debye model for the heat capacity of a solid ▷.

There are several important properties of the definite integral which are related to the limits and the behaviour of $f(x)$ in the interval $[a, b]$. For example, changing the order of the limits changes the sign of the integral:

$$\int_a^b f(x) \, dx = F(b) - F(a) = -\{F(a) - F(b)\} = -\int_b^a f(x) \, dx.$$

▷

That is $a < c < b$.

Problem 6.10

Show that for any point c lying inside the interval (a, b) ▷

$$\int_a^b f(x)\, dx = \int_a^c f(x)\, dx + \int_c^b f(x)\, dx.$$

Problem 6.11

For the following definite integrals, use an appropriate method for their evaluation. If the method of integration by parts is used, then the limits are handled in the usual way by evaluating the antiderivative function at the two end-points:

$$\int_a^b f(x) g'(x)\, dx = [f(x) g(x)]_a^b - \int_a^b f'(x) g(x)\, dx.$$

(a) $\displaystyle \int_2^7 \frac{1}{x^2}\, dx, \int_0^2 (1 + x)^3\, dx, \int_3^7 \frac{dx}{x^2 - 4}, \int_4^6 \frac{2x}{(x-3)(x-1)}\, dx.$

▷

See Problem 6.4.

(b) Show that ▷ $\displaystyle \int_0^2 \frac{x}{(x^2 + 4)}\, dx = \frac{1}{2}\ln 2.$

(c) $\displaystyle \int_0^L e^{-2r}\, dr, \int_0^L r^2 e^{-2r}\, dr,$ where L is a constant.

(d) $\displaystyle \int_{T_1}^{T_2} y\, dT,$ given that $y = a + bT + cT^2.$

Problem 6.12

If

$$f(x) = 4x^2 e^{-2x}$$

(a) find the two finite values of x for which the derivative of $f(x)$ vanishes, giving the respective values of $f(x)$ in each case;

(b) determine the second derivative of $f(x)$, and identify the nature of the two turning points found in part (a);

(c) sketch the graph of the function $f(x)$ for the subinterval $[-0.5, 2.0]$ of its domain;

(d) show that

$$\int f(x)\, dx = -e^{-2x}(2x^2 + 2x + 1) + C;$$

hence deduce that

$$\int_0^1 f(x)\,dx = 1 - 5/e^2;\tag{6.9}$$

(e) using the values of $(x, f(x))$ for the turning points found in part (a), and with reference to the graph of the function, show that $1 \cdot f(1)$ (the rectangular area formed by the points $(0,0), (1,0), (1,f(1)), (0,f(1)))$ and $0.5 \cdot f(1)$ (the triangular area formed by the points $(0,0), (1,0), (1,f(1)))$ form upper and lower limits, respectively, to the area under the curve. Thus deduce that $\sqrt{7} < e < 3$.

Problem 6.13

The form of the wavefunction for the particle in a one-dimensional box of length L was analysed in Problem 4.13. In this problem we wish to work out the average position, $<x>$, of the particle in the box, a result that is given by the integral $\triangleright$

▷
Alberty and Silbey,
Example 10.8.

$$<x> = \int_0^L \psi(x)\,x\,\psi(x)\;dx,$$

where

$$\psi(x) = \left(\frac{2}{L}\right)^{\frac{1}{2}} \sin\left(\frac{n\pi x}{L}\right).$$

▷
$\theta = n\pi x/L.$

Use the identity $1 - \cos 2\theta = 2\sin^2\theta$ ▷, to rewrite the integrand, and then use integration by parts to obtain the result $<x> = L/2$.

Problem 6.14

▷
Atkins, Equation (12a),
Section 9.4.

Let $K(T)$ be the equilibrium constant for formation of NH_3 from N_2 and H_2 at a given temperature, T. From thermodynamics, we know that ▷

$$\frac{d}{dT}\ln K(T) = \frac{\Delta H^\ominus}{RT^2}.$$

(a) Assuming that $\Delta H^\ominus$ is independent of temperature, determine how $\ln K(T)$ varies with T.

(b) Assuming now that $\Delta H^\ominus$ varies with temperature according to the expression

$$\Delta H^\ominus = a + bT + cT^2$$

derive an equation relating $\ln K(T)$ to T, and give the change in $\ln K(T)$ as the temperature changes from $T = T_1$ to $T = T_2$.

▷
Barrow, p. 652.

Problem 6.15

The *turbidity*, τ, associated with the scattering of light of wavelength, λ, from a macromolecule is given by the integral ▷

$$\tau = \int_0^\pi \frac{8\pi^4\alpha^2}{\lambda^4 r^2}(1+\cos^2\theta)\cdot 2\pi r^2 \sin\theta\,d\theta,$$

where α is the polarizability of the molecule.

Use the substitution $u = \cos\theta$ to show that

$$\tau = \frac{8\pi\alpha^2}{3}\left(\frac{2\pi}{\lambda}\right)^4.$$

6.7   Improper integrals

The result in Problem 6.10 demonstrates that it is possible to break down the range of the integration variable into contiguous subranges. This procedure is especially important in thermodynamic applications, where such breaks are placed at the temperatures where changes of physical phase lead to discontinuities in the function being integrated (see Figure 4.3). The methods used for the numerical evaluation of definite integrals also depend upon this principle of introducing contiguous subranges for the independent variable.

In fact, discontinuities and other convergence problems can occur at the end-points as well as within the interval of the independent variable over which the function is being integrated. The usual practice is to break the integration just before the troublesome point (and just after, if the point occurs inside the interval over which integration is taking place). Integrals with infinite upper or lower limits, or with discontinuities arising within $[a, b]$, are formally termed *improper integrals*. The behaviour of the integrand in the neighbourhood of suspect points in $[a, b]$ is probed by using the kinds of limiting process discussed in Chapter 3. Illustrations of the procedure are given in the following example.

Worked example

6.4   (a) Evaluate $\displaystyle\int_0^1 \frac{1}{(1-x)^{\frac{1}{2}}}\,dx$.

Solution The integrand has a discontinuity at $x = 1$, hence the attempt to evaluate this integral must proceed by first evaluating the integral

$$\int_0^{1-\epsilon} \frac{1}{(1-x)^{\frac{1}{2}}}\, dx,$$

where, now, the discontinuity is removed. The resulting value for this new integral is a function of ϵ, and we then proceed to take the limit of this function as $\epsilon \to 0$. If the limit exists, then the integral is said to converge, and the value obtained is the value of the original integral; if the limit does not exist, then the original integral is said to diverge, and it has no value. In the present example, the modified integral is most easily evaluated by using the substitution $u = (1-x)$ ▷:

$\triangleright$

Notice that the limits also change with the change of variable since, when $x = 0$, and $1 - \epsilon$, $u = 1$ and ϵ, respectively.

$$\int_0^{1-\epsilon} \frac{1}{(1-x)^{\frac{1}{2}}}\, dx = -\int_1^{\epsilon} u^{-\frac{1}{2}}\, du = -[2u^{\frac{1}{2}}]_1^{\epsilon} = 2 - 2\epsilon^{\frac{1}{2}}.$$

Thus,

$$\lim_{\epsilon \to 0} \int_0^{1-\epsilon} \frac{1}{(1-x)^{\frac{1}{2}}}\, dx = \lim_{\epsilon \to 0}(2 - 2\epsilon^{\frac{1}{2}}) = 2,$$

and the integral converges.

(b) Evaluate $\int_0^1 \frac{1}{x^3}\, dx.$

Solution The integrand has an infinite discontinuity at $x = 0$, the lower limit. Thus

$$\int_\epsilon^1 \frac{1}{x^3}\, dx = \left[-\frac{1}{2}x^{-2}\right]_\epsilon^1 = -1 + \frac{1}{\epsilon^2},$$

which, in the limit $\epsilon \to 0$ becomes infinite – thereby indicating that the integral diverges and has no value.

(c) Evaluate $I = \int_0^5 \frac{1}{x^2 - 9}\, dx.$

Solution Here there is a discontinuity at $x = 3$, which lies inside the interval over which the integration is being made. The first step involves carrying out the integration with the point $x = 3$ missing. If the method of partial fractions is used first, then

$$I = \int_0^5 \frac{1}{x^2 - 9}\, dx = \frac{1}{6}\int_0^5 \left\{\frac{1}{x-3} - \frac{1}{x+3}\right\} dx$$

$\triangleright$

$\epsilon > 0.$

which, avoiding the discontinuity arising from the first term, becomes ▷

$$I = \frac{1}{6}\lim_{\epsilon \to 0}\int_0^{3-\epsilon} \frac{1}{x-3}\, dx + \frac{1}{6}\lim_{\epsilon \to 0}\int_{3+\epsilon}^5 \frac{1}{x-3}\, dx - \frac{1}{6}\int_0^5 \frac{1}{x+3}\, dx.$$

The resulting three integrals, evaluated using the substitution procedure, all yield logarithmic functions:

$$I = \frac{1}{6} \lim_{\epsilon \to 0} [\ln(|x - 3|)]_0^{3-\epsilon} + \frac{1}{6} \lim_{\epsilon \to 0} [\ln(|x - 3|)]_{3+\epsilon}^5 - \frac{1}{6} [\ln(|x + 3|)]_0^5$$

$$= \frac{1}{6} \lim_{\epsilon \to 0} [\ln \epsilon - \ln 3] + \frac{1}{6} \lim_{\epsilon \to 0} [\ln 2 - \ln \epsilon] - \frac{1}{6} [\ln 8 - \ln 3].$$

The first two integrals diverge in the limit as $\epsilon \to 0$, as the logarithm of a decreasingly small positive number increases without limit in a negative sense. However, if we choose ϵ in each limit to take the same value as the limit is approached then the terms cancel, yielding the finite value

▷
The value obtained is termed the Cauchy principal value.

$$\frac{1}{6} \{ -\ln 3 + \ln 2 \}, ▷$$

hence the overall value for I is $-\frac{1}{3} \ln 2$.

(d) Evaluate $\int_0^\infty e^{-2r} \, dr$.

Solution   Since the upper limit is infinite, this form of improper integral must be evaluated by first replacing the upper limit by a large number, L say, and then taking the limit as $L \to \infty$: that is

$$\lim_{L \to \infty} \int_0^L e^{-2r} dr.$$

The integral with finite limits is now in standard form, and the result is $[-e^{-2r}/2]_0^L = (1 - e^{-2L})/2$. Thus,

$$\lim_{L \to \infty} \int_0^L e^{-2r} dr = \lim_{L \to \infty} (1 - e^{-2L})/2 = 1,$$

and the integral converges.

Problem 6.16

Evaluate

(a) $\int_0^7 \dfrac{dx}{x^2 - 4}$ (b) $\int_0^\infty r^2 e^{-2r} dr$,

▷
See Problem 6.3(e).

(c) ▷ $\int_{-\infty}^\infty \dfrac{dx}{(a^2 + x^2)^{52}} = \lim_{L \to \infty} \int_{-L}^L \dfrac{dx}{(a^2 + x^2)^{52}}$.

▷

Barrow, p. 107; note the misprint in Alberty and Silbey, Equation (18.20).

Problem 6.17

The average speed, $<v>$, of a molecule of mass, m, in a gaseous sample at temperature, T, is given by the model provided by the kinetic theory of gases ▷ as the expression

$$<v> = 4\pi \left(\frac{m}{2\pi kT}\right)^{\frac{3}{2}} \int_0^\infty e^{-\alpha v^2} v^3 \, dv,$$

where k is the Boltzmann constant, and $\alpha = m/(2kT)$. Use
(a) the substitution $u = v^2$ to simplify the integral, and

(b) the method of integration by parts to show that $<v> = \left(\frac{8RT}{\pi M}\right)^{\frac{1}{2}}$, where M is the molar mass.

▷

Alberty and Silbey, Section 17.9.

▷

Alberty and Silbey, Section 22.4; note that the integrand in their Equation (22.6) needs an extra r^2 term.

Problem 6.18

(a) The rotational partition function, q_r, which involves a summation over rotational states, is needed for computing the rotational contribution to thermodynamic properties of gases, such as heat capacity, entropy, free energy, etc. Using the observation that, because the energy states are closely spaced for molecules with small values of Θ_r (large moments of inertia) the summation over states is well approximated by the integral ▷:

$$q_r = \int_0^\infty (2J + 1) \, e^{-J(J+1)\lambda} \, dJ,$$

where $\lambda = \Theta_r/T$. Use the substitution $x = J(J+1)$ to evaluate this integral.

(b) In the freely jointed model of a long chain polymer of length l, with n links, the average of the square of the end-to-end distance, $<r^2>$, is given by ▷

$$<r^2> = 4\pi \left(\frac{3}{2\pi n l^2}\right)^{\frac{3}{2}} \int_0^\infty e^{-\frac{3r^2}{2nl^2}} r^4 \, dr.$$

Use the result

$$\int_0^\infty u^4 e^{-u^2} \, du = \frac{3}{2}\sqrt{\pi}/4$$

to show that $<r^2> = nl^2$.

6.8   Numerical determination of definite integrals

As noted earlier, it is always possible in principle to evaluate definite integrals numerically. For integrals other than improper integrals there is no real problem if the sign of $f(x)$ does not change in the interval $[a, b]$. However, if $f(x)$ oscillates in sign, then care must be exercised to avoid the loss of precision in the cumulative addition of positive and negative contributions to the definite integral. It is advantageous in such a situation to locate the zeros in the integrand and then break the interval $[a, b]$ at these values of x.

Improper definite integrals pose additional problems, in that the range of integration must be broken at points of finite discontinuity in the interval $[a, b]$; also points of infinite discontinuity, and infinite values for the end-points a or b must also be recognized and treated accordingly ▷. Fortunately, in a chemical context, the definite integrals that arise relate to physical observables, and will therefore exist! This does not mean, however, that a numerical determination of the value for the integral will be devoid of problems.

There are many ways of approaching the numerical determination of definite integrals. For purposes of illustration, we shall just examine the *trapezoidal* rule, where the interval $[a, b]$ is subdivided into n equal subintervals of width h, a very simple application of which was seen in Problem 6.12, where only one subinterval was used in order to obtain some bounds on e.

Using the trapezoidal rule, the $n + 1$ values of x, designated by $x_0, x_1, \ldots, x_n$ are associated with the values of $f(x)$ given by $f(x_0), f(x_1), \ldots, f(x_n)$. Neighbouring points $(x_i, f(x_i)), (x_{i+1}, f(x_{i+1}))$ are joined by straight line segments, so that a trapezium stands on the subinterval of width $h = (b - a)/n = x_{i+1} - x_i$ $(i = 0, 1, \ldots, n - 1)$ (see Figure 6.3). Now the area of each such trapezium is the sum of the areas of the rectangle and triangle as shown in Figure 6.3: that is,

$hf(x_i) + h\{f(x_{i+1}) - f(x_i)\}/2 = h\{f(x_{i+1}) - f(x_i)\}/2$, which is half the sum

▷

See, for example Press *et al.*, Chapter 4.

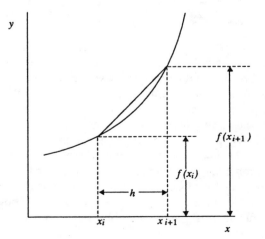

Figure 6.3
Construction of a trapezium on the subinterval between x_i and x_{i+1} of width h

of the parallel sides times the distance between them. The value, I, for the definite integral can thus be written as

$$I = \sum_{i=0}^{n-1} \frac{h}{2}(f(x_{i+1}) + f(x_i)) + E$$

$$= \frac{h}{2}f(x_0) + hf(x_1) + hf(x_2) + \cdots + hf(x_{n-1}) + \frac{h}{2}f(x_n) + E$$

$$= \sum_{i=0}^{n-1} w_i f(x_i) + E,$$

▷

$w_0 = w_{n-1} = h/2$,
$w_i \ (i \neq 0, n-1) = h$.

where w_i is a weight factor ▷, and E is the error in approximating I using the trapezoidal rule. Clearly, increasing the number of subintervals reduces E, but if n becomes too large then precision may be lost on account of the increasing number of arithmetical operations with associated rounding errors. Other numerical schemes involve using different sets of x_i and w_i with or without the transformation of the range of integration from $[a, b]$ to $[-1, 1]$ ▷. Since different numerical algorithms yield different error terms, E, it is important to select an algorithm to make E as small as desired.

▷

Press *et al.*, or Stroud and Secrest.

> ## Problem 6.19
>
> The specific heat at constant pressure, C_p, for solid benzene is as follows for $23 \text{ K} \leq T \leq 273 \text{ K}$:
>
> | T/K | 23 | 48 | 73 | 123 | 223 | 273 |
> |---|---|---|---|---|---|---|
> | $C_p/\text{J mol}^{-1}\text{K}^{-1}$ | 13.01 | 29.66 | 40.46 | 55.44 | 97.61 | 122.30 |
>
> Assuming that $C_p = AT^3$, for $0 \text{ K} \leq T \leq 23 \text{ K}$, determine the entropy of solid benzene at 273 K by evaluating the definite integral $\int_0^{273} \frac{C_p}{T} dT$ using the trapezoidal rule, and splitting the range of integration at 23 K.

Summary: This chapter introduces the idea of an operator for reversing the effects of differentiation when dealing with a function of a single variable. There then follow sections covering different techniques for transforming a given integral into a more recognizable form – be it a definite or an indefinite integral. It is stressed that integration, unlike differentiation, is much less of a mechanical process and it may not always be possible to determine an antiderivative function in closed form. The chapter concludes with a brief introduction to the use of numerical methods for evaluating definite integrals. Throughout the chapter, the discussion and development of the material is illustrated with examples involving chemical applications whenever possible.

The next chapter develops the concept of power series representations of a function of a single variable, and incorporates the techniques of both the differential and the integral calculus.

7 Power series: a new look at functions

> This chapter shows how
>
> - the value of a function and its derivatives at a given point are used to determine the value of the function at a neighbouring point
> - the domain of the power series representation of the function may be determined
> - power series representations of functions may be differentiated and integrated
> - power series arise within a chemical context

▷

In principle, an indefinite number of terms.

We have already seen in Chapter 2 that there are certain situations in chemistry where we just do not know the algebraic form of a function. For example, the prescription that delineates how the enthalpy change of a process varies with temperature is commonly written as a sum of terms ▷ in the form $\Delta H = a_0 + a_1 T + a_2 T^2 + \cdots$, where successive terms involve an increasing power of the independent variable, T. This representation of a function is termed a power series. In the present example, where only the first few terms are usually considered, the coefficients, a_r, are determined by fitting the formula to experimental data.

▷

Arising from the averaged effects of the electronic motion and the repulsive interaction of the nuclei.

Another example of a power series is seen in the modelling of the vibrational potential energy function ▷ for describing the motion of the nuclei in a molecule. For a diatomic molecule the form of the potential energy function is usually expressed in terms of a power series in the bond displacement coordinate as measured from its equilibrium value. Just as in the thermodynamic example, however, the series is usually terminated after a few terms – an approximation that limits the applicability of the model to studying excited vibrational states lying well below the dissociation limit.

It should be noted that, within the context of chemistry, not all series occur in power series form. In crystalline sodium chloride, for example, with a nearest neighbour distance of R, the potential, V, experienced at a sodium site is calculated by summing the infinite number of contributions from neighbouring sodium and chlorine ions at progressively increasing distance from the reference sodium ion (ionic model assumed):

$$V = \frac{1}{(4\pi\epsilon_0 R)}\{c_0 + c_1 + c_2 + \cdots\} = a_0 + a_1 + a_2 + \cdots .$$

The series that arises here does not contain powers of the independent variable, and its sum can be expressed in terms of the Madelung constant, A (depending only on the geometry of the crystal structure), and R ▷. Thus, for this and other similar series of constant terms, the function prescription contains only a fixed power of the independent variable. We are concerned here with power series, in which successive terms depend on a different power of the independent variable.

▷
Smart and Moore, p. 52.

7.1	The Maclaurin series

The power series representation of a function is written as a sum of terms, each with a coefficient a_r $(r = 0, 1, 2, \ldots)$ multiplying the independent variable raised to the rth power:

$$f(x) = a_0 + a_1 x + a_2 x^2 + \cdots + a_r x^r + \cdots ,$$

where $\cdots$ signifies that the series continues indefinitely. The rth term is given by $a_{r-1}x^{r-1}$. Using the summation convention, this series takes the more compact form

$$\sum_{k=0}^{\infty} a_k x^k ,$$

where k is a counting (or dummy) index ▷. A power series in this form is termed a *Maclaurin series*, or a Maclaurin expansion of $f(x)$, and the method for determining the coefficients a_r for given choices of $f(x)$ is described below.

▷
There is nothing special about the name k – it could be $r, s, A,$ etc.

Testing for convergence

The techniques that are deployed in determining whether a particular series sums to finite values (converges) for selected values of x are very subtle. It so happens, however, that the convergence of power series, and the associated problem of determining the values of x for which successive terms $u_r = a_{r-1}x^{r-1}$ of such series do sum to a finite value as $r \to \infty$, is a simpler problem as we can test for *absolute* convergence ▷. Fortunately, a more detailed mathematical study than we can offer here shows that absolutely convergent series are also convergent: a result we accept without proof. The

▷
Essentially this means that we are examining the convergence of a series of positive terms.

test for absolute convergence is carried out using the *ratio* test. If u_r and u_{r+1} are the rth and $(r+1)$th terms in the Maclaurin series, then the series converges if

$$\lim_{r\to\infty}\left|\frac{u_{r+1}}{u_r}\right| < 1.$$

▷

An important point to notice is that the domain determined for a power series will, in general, be a subset of the domain of its parent function.

The application of this test yields a set of x values forming the *interval of convergence* (the domain) for the power series representation of the function ▷.

Worked example

7.1   Find the values of x for which the series

$$1 + x + x^2 + x^3 + \cdots + x^r + \cdots \tag{7.1}$$

converges.

Solution   The rth and $(r+1)$th terms are given by $u_r = x^{r-1}$ and $u_{r+1} = x^r$, respectively. Thus,

$$\lim_{r\to\infty}\left|\frac{u_{r+1}}{u_r}\right| = \lim_{r\to\infty}\left|\frac{x^r}{x^{r-1}}\right| = \lim_{r\to\infty}|x| = |x|,$$

and the third equality arises because the expression $|x|$ in the limit does not depend upon r, and therefore remains constant in the limit as $r \to \infty$. The series therefore converges so long as $|x| < 1$; that is, $-1 < x < 1$. The question as to whether the series converges at either of the end-points in the interval of convergence is usually a delicate matter to resolve and, in this case, the series diverges (does not sum to a finite value) at both end-points. At $x = 1$, the sum of the first r terms is r, and this increases without limit as $r \to \infty$ but, at $x = -1$, the sum of the series oscillates between zero and one, depending whether r is even or odd, and a finite value for the sum is not obtained as $r \to \infty$.

Problem 7.1

Consider the sum of the first $(r+1)$ terms of the series in the previous example (a *geometric progression*):

$$S_r = 1 + x + x^2 + \cdots + x^r = \sum_{k=0}^{r} x^k.$$

(a) Write down xS_r, and hence show that, by subtracting the two series, the following expression for the sum of the series is obtained:

$$S_r = \left(\frac{1 - x^{r+1}}{1 - x}\right).$$

(b) Use the result in Example 7.1, which shows that the infinite series converges only for $|x| < 1$, to demonstrate that

$$\lim_{r \to \infty} S_r = \sum_{k=0}^{\infty} x^k = \frac{1}{1-x} .$$

(c) Give an expression for the sum of the series $1 + \dfrac{1}{2} + \dfrac{1}{2^2} + \dfrac{1}{2^3} + \cdots$

and $1 + e^\theta + e^{2\theta} + e^{3\theta} + \cdots \triangleright$.

$\triangleright$
θ such that $e^\theta < 1$.

$\triangleright$
That is, the domain is
$\mathbb{R} - \{1\}$.

The results obtained in Example 7.1 and Problem 7.1 demonstrate a point hinted at earlier, that the function $f(x) = (1-x)^{-1}$ and its Maclaurin power series representation in Equation (7.1) have different domains. The function formula for $f(x)$ is valid for all x except $x = 1$ $\triangleright$, whilst the associated Maclaurin series only converges for x lying in the subset $-1 < x < 1$, often written in the form $(-1, 1)$.

7.2   The Taylor series

So far we have examined one series in powers of x (the Maclaurin form), and discovered that it was related to the function $(1-x)^{-1}$. However, if $f(x)$ is expanded in terms of powers of $(x - a)$, where a is a fixed number, then the Maclaurin expansion is a special case with $a = 0$ (expansion about the origin). In this *Taylor* series form of the power series expansion, there is more flexibility in the choice of the point a about which to expand $f(x)$. Different power series representations of $f(x)$ are obtained, depending upon the choice of a. Each such series has a different interval of convergence (domain).

Let us now return to the problem of finding Taylor series representations of a given function, where $f(x)$ is expanded in powers of $(x - a)$ in the form

$$f(x) = c_0 + c_1(x - a) + c_2(x - a)^2 + \cdots + c_r(x - a)^r + \cdots$$
$$= \sum_{k=0}^{\infty} c_k(x - a)^k . \tag{7.2}$$

The constants $c_0, c_1, c_2, \ldots$, which have different values from the corresponding coefficients in the Maclaurin series, depend upon the choice of a and, of course, $f(x)$.

The values of the coefficients c_k are found in the following sequence of steps:

(1) Substitute $x = a$ in Equation (7.2) to yield $c_0 = f(a)$, as all terms in the series except the first one are zero.

(2) Differentiate both sides of Equation (7.2) to obtain the expression

$$f^{(1)}(x) = c_1 + 2c_2(x - a) + 3c_3(x - a)^2 + \cdots + rc_r(x - a)^{r-1} + \cdots$$

$$(7.3)$$

and substitute $x = a$, to obtain $c_1 = f^{(1)}(a)$.

(3) Differentiate Equation (7.3) twice, evaluating successive derivatives at $x = a$, to obtain

$$f^{(2)}(a) = 1 \cdot 2c_2 \Rightarrow c_2 = \frac{f^{(2)}(a)}{2!}$$

$$f^{(3)}(a) = 1 \cdot 2 \cdot 3c_3 \Rightarrow c_3 = \frac{f^{(3)}(a)}{3!}.$$

At this juncture, a pattern is observed in the expressions for the derivatives and the related coefficients, thus enabling us to write down the rth derivative as $f^{(r)}(a) = 1 \cdot 2 \cdot \ldots r\, c_r \Rightarrow c_r = \frac{f^{(r)}(a)}{r!}$.

Hence, substituting the values for each of the coefficients, c_r, in the power series representation for $f(x)$ enables us to write the value of a function at the point x in terms of the values of the function and its derivatives at the point a as

$$f(x) = f(a) + f^{(1)}(a)(x - a) + \frac{f^{(2)}(a)}{2!}(x - a)^2 + \cdots$$

$$+ \frac{f^{(r)}(a)}{r!}(x - a)^r + \cdots.$$

$$(7.4)$$

This form for $f(x)$ should be compared with Equation (5.1).

The interval of convergence for the Taylor expansion of $f(x)$ is determined by applying the ratio test:

$$\lim_{r \to \infty} \left| \frac{u_{r+1}}{u_r} \right| = \lim_{r \to \infty} \left| \frac{f^{(r)}(a)(x - a)^r}{r!} \times \frac{(r - 1)!}{f^{(r-1)}(a)(x - a)^{r-1}} \right|$$

$$= \lim_{r \to \infty} \left| \frac{f^{(r)}(a)(x - a)}{rf^{(r-1)}(a)} \right|.$$

For convergence, the last limit must yield a value which is less than unity, and this requires specific knowledge of the form of $f(x)$. If $f(x)$ is evaluated for x lying within the interval of convergence for the power series, then it is possible to terminate the series at a particular term (say the $(p + 1)$th), when the value of $f(x)$ has converged to the required number of decimal places. The (infinite) power series is now reduced to the simpler Taylor polynomial function

$$f(a) + \cdots + \frac{f^{(p)}(a)}{p!} \cdot (x - a)^p,$$

$$(7.5)$$

of degree p with an error given by the difference between the actual value of $f(x)$ and the value obtained from the Taylor polynomial. Clearly, if we want to maintain the same error for different values of x, it is necessary to use a different Taylor polynomial for each value of x.

$\triangleright$

$\Delta x = x - a \equiv h.$

A comparison with the analysis associated with the differential in Chapter 5 shows that the difference between dy and Δy, previously written as $\epsilon \Delta x \, \triangleright$, is given explicitly by the third and subsequent terms in the Taylor series, Equation (7.4).

Worked example

$\triangleright$

See Equation (7.5).

7.2   (a) Find the Taylor series representations of the function $f(x) = (2 + x)^{-1}$, for $a = 0$ and $a = 1$.

(b) Determine their respective intervals of convergence (domains).

(c) Compare the exact value for $f(0.9)$ with the values obtained using the Taylor polynomials $\triangleright$ of degree 1 and 3, giving answers to four decimal places. (This means working to five decimal places and rounding the answer.)

Solution   (a) The Taylor series requires the determination of the value of the function and its derivatives at an arbitrary point a:

$$f(x) = (2 + x)^{-1} \Rightarrow f(a) = \frac{1}{(2 + a)}$$

$$f^{(1)}(x) = (-1)(2 + x)^{-2} \Rightarrow f^{(1)}(a) = (-1)\frac{1}{(2 + a)^2}$$

$$f^{(2)}(x) = (-1)(-2)(2 + x)^{-3} \Rightarrow f^{(2)}(a) = (-1)^2\frac{2!}{(2 + a)^3}$$

$$f^{(3)}(x) = (-1)^3 \cdot 1 \cdot 2 \cdot 3(2 + x)^{-4} \Rightarrow f^{(3)}(a) = (-1)^3\frac{3!}{(2 + a)^4}.$$

In formulating the expressions for successive derivatives in the manner shown above, it becomes possible to recognize the pattern, and thereby write down the rth derivative (the coefficient of $(x - a)^r$ in the $(r + 1)$th term). It is a good idea to check the expression by substituting values for r, noting that the case $r = 0$ does not always emerge from the general formula.

In this case, it is easy to see that

$$f^{(r)}(x) = (-1)^r \cdot 1 \cdot 2 \cdot 3 \cdots r(2 + x)^{-(r+1)} \Rightarrow$$

$$f^{(r)}(a) = (-1)^r \frac{r!}{(2 + a)^{r+1}}.$$

Thus, the general Taylor series for the function $f(x) = (2+x)^{-1}$ is

$$\frac{1}{(2+a)} - \frac{1}{(2+a)^2}(x-a) + \frac{1}{(2+a)^3}(x-a)^2 + \cdots$$
$$+ \frac{(-1)^r}{(2+a)^{r+1}}(x-a)^r + \cdots,$$

which simplifies to

$$\frac{1}{2+a}\left\{1 - \left(\frac{x-a}{2+a}\right) + \left(\frac{x-a}{2+a}\right)^2 - \left(\frac{x-a}{2+a}\right)^3 + \cdots\right.$$
$$\left. + (-1)^r\left(\frac{x-a}{2+a}\right)^r + \cdots\right\}.$$

Thus, the Taylor series for $a=0$ and $a=1$ become

$$(2+x)^{-1} = \frac{1}{2}\left\{1 - \left(\frac{x}{2}\right) + \left(\frac{x}{2}\right)^2 - \left(\frac{x}{2}\right)^3 + \cdots\right.$$
$$\left. + (-1)^r\left(\frac{x}{2}\right)^r + \cdots\right\}, \tag{7.6}$$

and

$$(2+x)^{-1} = \frac{1}{3}\left\{1 - \left(\frac{x-1}{3}\right) + \left(\frac{x-1}{3}\right)^2 - \left(\frac{x-1}{3}\right)^3 + \cdots\right.$$
$$\left. + (-1)^r\left(\frac{x-1}{3}\right)^r + \cdots\right\}, \tag{7.7}$$

respectively.

(b) The application of the ratio test to find the respective interval of convergence requires the examination of the behaviour of the magnitude of the ratio of the $(r+1)$th and rth terms as r becomes indefinitely large:

$$u_r = \frac{(-1)^{r-1}}{2+a}\left(\frac{x-a}{2+a}\right)^{r-1}; \quad u_{r+1} = \frac{(-1)^r}{2+a}\left(\frac{x-a}{2+a}\right)^r.$$

Thus

$$\lim_{r\to\infty}\left|\frac{u_{r+1}}{u_r}\right| = \left|\frac{x-a}{2+a}\right|$$

and the series converges if

$$\left|\frac{x-a}{2+a}\right| < 1.$$

Thus, since multiplying both sides of the inequality by the positive number $|2 + a|$ does not change its sense, it follows that

$$|x - a| < |2 + a| \Rightarrow -|2 + a| < x - a < |2 + a|$$
$$\Rightarrow a - |2 + a| < x < a + |2 + a|.$$

▷
Or $(-2, 2)$ and $(-2, 4)$,
respectively (we ignore
what happens at the end-
points).

Thus the intervals of convergence are $-2 < x < 2$ for $a = 0$, and $-2 < x < 4$ for $a = 1$ ▷.

(c) The calculation of $f(0.9)$ is carried out using the expressions given in the second step. The exact value of $f(0.9)$ is $(2 + 0.9)^{-1} = 0.3448$, to four decimal places. The first two terms of the Maclaurin series ($a = 0$) yield a value of 0.275, whilst the first four terms yield a value of 0.3307. Similarly, the corresponding terms for the Taylor series with $a = 1$ are 0.3444 and 0.3448, respectively.

7.3   Manipulating power series

The results in Example 7.2 illustrate two important points. Firstly, an appropriate choice of expansion point a can result in a good polynomial representation of the infinite series, so long as a is close to the value of x for which $f(x)$ is required; secondly, the intervals of convergence (domains) of different power series expansions of a function are different (in the present example, the expansion for $a = 1$ leads to a larger interval of convergence than that for $a = 0$). The implications of these observations are examined further in Problem 7.11, where different polynomial expansions of the Morse potential ▷ for a diatomic molecule are examined.

▷
Used in determining
vibrational energies.

The previously derived series, given in Equation (7.6), may be modified in a simple way to provide Maclaurin series for related functions. For example, as

$$\frac{1}{2 + x} = \frac{1}{2(1 + x/2)} = \frac{1}{2}\left\{1 - \left(\frac{x}{2}\right) + \left(\frac{x}{2}\right)^2 + \cdots + (-1)^r\left(\frac{x}{2}\right)^r + \cdots\right\},$$

the series for $g(x) = (1 + x/2)^{-1}$ follows immediately in the form

$$g(x) = \left\{1 - \left(\frac{x}{2}\right) + \left(\frac{x}{2}\right)^2 - \left(\frac{x}{2}\right)^3 + \cdots + (-1)^r\left(\frac{x}{2}\right)^r + \cdots\right\}.$$

Noting that if we write $w = x/2$ then,

$$g(2w) = h(w) = (1 + w)^{-1}$$
$$= \{1 - w + w^2 - w^3 + \cdots + (-1)^r w^r + \cdots\}.$$

Since there is nothing special with the variable name used for specifying the function rule, we can write x instead of w, and obtain the result

$$h(x) = (1 + x)^{-1} = \{1 - x + x^2 - x^3 + \cdots + (-1)^r x^r + \cdots\}. \qquad (7.8)$$

Thus, the Maclaurin series for the function $(1 + x)^{-1}$ is related directly to the series for $(2 + x)^{-1}$.

Problem 7.2

(a) Derive the Maclaurin series for $(1 - x)^{-1}$ using Equation (7.8).

(b) The vibrational partition function for a diatomic molecule, needed for the calculation of thermodynamic properties, is given by ▷

$$q_v = \sum_{m=0}^{\infty} e^{-mhv/kT}.$$

By making the substitution $x = e^{-hv/kT}$, demonstrate that ▷

$$q_v = \frac{1}{1 - e^{-\Theta_v/T}},$$

where $\Theta_v = hv/k$ is the characteristic vibrational temperature.

(c) Use the result in part (a), together with the procedure used in determining the Maclaurin expansion for $(2 + x)^{-1}$ ▷ to obtain the Maclaurin series for $(3 - 2x)^{-1}$.

▷

Alberty and Silbey,
Section 17.8.

▷

See Problem 7.1(c).

▷

See the earlier part of
this section.

Problem 7.3

Find the Taylor series for the function $f(x) = (2 + x)^{-1}$ about the point $a = 2$, and determine its interval of convergence. What is the degree of the Taylor polynomial that yields $f(0.9)$ to four decimal places?

Problem 7.4

Find the Taylor expansion for the function $f(x) = x^3 + 3x + 1$ about the point $a = 1$ ▷.

▷

In this case a Taylor
polynomial is obtained.

Problem 7.5

Use the results of Problem 4.6 to derive the Maclaurin expansion for each of the functions

$$e^x, \ e^{-x}, \ \cos x, \ \sin x, \ \ln(1 + x).$$

In each case give the rth term, and apply the ratio test to determine the values of x for which the series converges ▷.

▷

Remember that
$(n + r)!/n! =$
$(n + r)(n + r - 1) \cdots$
$(n + 1)$, and that
$(n + 1)/n$ can be written
as $1 + 1/n$.

Problem 7.6

Use the Maclaurin expansion for e^x to determine the degree of the Taylor polynomial that yields $e^{0.1}$ to three places of decimals; repeat the calculation for $e^{0.2}$.

Problem 7.7

(a) Verify the assertion in Chapter 4, in the derivation of $\lim_{\theta \to 0} \sin \theta / \theta$, that the second-degree polynomial approximation to the Maclaurin series for $\sec x$ is $1 + x^2$.

(b) Write down the Maclaurin series for $- \cos x$ using the answer for $\cos x$ in Problem 7.5 above; hence, find the Maclaurin series for $\sin x$ by differentiation, taking care to identify the rth term ▷.

▷ Only non-zero terms are counted.

(c) Differentiate the Maclaurin series for $\ln (1 + x)$ in Problem 7.5 and, from Equation (7.8), identify the function possessing the series so obtained.

(d) Integrate the Maclaurin series for $\ln (1 + x)$ and give its interval of convergence.

▷ Take $f(x) = \ln (1 + x)$ and $g'(x) = 1$.

(e) Integrate the function $\ln (1 + x)$ using integration by parts ▷ to obtain the function whose Maclaurin series is given in part (d).

Problem 7.8

▷ Alberty and Silbey, Section 22.7; x is used here instead of p.

The number average molar mass, $\overline{M}_n$, and the mass average molar mass, $\overline{M}_m$, used to characterize a growing polymer, are given by ▷

$$\overline{M}_n = M_0(1 - x) \sum_{k=1}^{\infty} k\, x^{k-1} \text{ and } \overline{M}_m = M_0 \frac{\sum_{k=1}^{\infty} k^2\, x^{k-1}}{\sum_{k=1}^{\infty} k\, x^{k-1}},$$

where x is equal to the fraction of monomers reacted, and M_0 is the molar mass of the monomer. Show that

$$\overline{M}_n = \frac{M_0}{(1 - x)} \text{ ; and } \overline{M}_m = M_0 \left(\frac{1 + x}{1 - x} \right).$$

The solution of this problem is best approached in a number of well-defined steps.

(a) Start with the Maclaurin series of the function $(1 - x)^{-1}$ which was obtained as the solution to Problem 7.2(a):

$$1 + x + x^2 + x^3 + \cdots + x^r + \cdots = \sum_{k=0}^{\infty} x^k.$$

(b) Differentiate the function $(1 - x)^{-1}$, and its Maclaurin series term by term, and show that the series representation of $(1 - x)^{-2}$ may be written as

$$(1 - x)^{-2} = 1 + 2x + 3x^2 + 4x^3 + \cdots + rx^{r-1} + \cdots$$

$$= \sum_{k=1}^{\infty} k\, x^{k-1}. \tag{7.9}$$

(c) Differentiate both sides of Equation (7.9) again, and deduce that

$$2(1 - x)^{-3} = 2 + 3 \cdot 2x + 4 \cdot 3x^2 + \cdots + r(r - 1)x^{r-2} + \cdots$$

$$= \sum_{k=2}^{\infty} k(k - 1)\, x^{k-2}.$$

(d) Change the counting index by letting $k = k' + 1$, and then drop the prime to show that

$$2(1 - x)^{-3} = \sum_{k=1}^{\infty} k^2 x^{k-1} + \sum_{k=1}^{\infty} k\, x^{k-1}. \tag{7.10}$$

(e) Use Equation (7.9) to substitute for the last summation in Equation (7.10) and show that

$$\sum_{k=1}^{\infty} k^2 x^{k-1} = \frac{1 + x}{(1 - x)^3}. \tag{7.11}$$

(f) Substitute Equations (7.9) and (7.11) into the given expressions for $\overline{M}_m$ and $\overline{M}_n$ to establish the results

$$\overline{M}_n = \frac{M_0}{(1 - x)} \quad \text{and} \quad \overline{M}_m = M_0\left(\frac{1 + x}{1 - x}\right).$$

The last part of Problem 7.7 shows that it is not always an easy job to recognize the function giving rise to a power series that has been obtained by differentiating or integrating another power series – unless, of course, the differentiation or integration is also carried out on the function itself.

Other manipulations of power series, such as addition, multiplication and division, are defined if the appropriate (binary) operation is carried out only for those values of x which lie in the interval of convergence of both series. For example, it is meaningless to add a series defined in $(-1, 1)$ to another series defined in $(2, 6)$ since the domains are disjoint – that is, there is no value of x lying in both domains. However, a series defined in $(-1, 1)$ can be added to another series defined in $(0, 1)$, say, to obtain a series which yields the sum of the two series in $(0, 1)$ ▷.

▷
A shorthand for the interval $0 < x < 1$; see also Example 2.4.

Problem 7.9

Give the Maclaurin expansions, including the rth terms, for the functions

(a) $x^2 e^x$,   (b) $\left(\dfrac{1+x}{1-x}\right)$,   (c) $\dfrac{e^x}{1-x}$.

Problem 7.10

(a) Use the procedure given in Example 7.2(a) to show that the Maclaurin series for the function $f(x) = (1+x)^\alpha$ can be written as

$$f(x) = \sum_{r=0}^{\infty} \binom{\alpha}{r} x^r$$

where

$$\binom{\alpha}{r} = \frac{\alpha(\alpha-1)\cdots(\alpha-r+1)}{1\cdot 2\cdot 3\cdots r}, \text{ and } \binom{\alpha}{0} = 1.$$

Notice that the upper limit of the sum is infinite except when α is a positive integer and then, as already noted in Problem 7.4, $f(x)$ is a polynomial function of degree α.

(b) Determine the interval of convergence of the Maclaurin series for $f(x)$.

(c) Give the Maclaurin series for the cases $\alpha = -1$ and $\alpha = \frac{1}{2}$.

(d) Evaluate $(1.01)^{-1}$ and $(1.1)^{1/2}$ to three decimal places.

Problem 7.11

Consider a diatomic molecule A–B. The molecular potential energy curve shows a variation of energy, $E(R)$, with internuclear separation, R, and displays a minimum at $R = R_e$ as seen in Figure 7.1. The energy attains an asymptotic value as R tends to infinity, corresponding to the formation of separated atoms.

(a) Write down the Taylor series for $E(R)$ about the point R_e.

(b) What is the value of $E'(R_e)$?

(c) Write down the Taylor polynomials of degree two and three in terms of the derivatives of $E(R)$ evaluated at the point R_e.

(d) Use your result in (c) to write down the expression for the energy, $\mathcal{E}(R) = E(R) - E(R_e)$, to second order when $E(R_e)$ is taken as the zero of energy; give a rough sketch of the function in $\mathcal{E}$ in the vicinity of R_e.

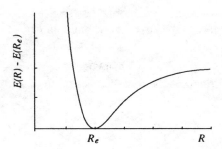

Figure 7.1
The variation in the
potential energy, $E(R)$, for
the vibration of a diatomic
molecule as a function of
the internuclear separation,
R. $E(R_e)$ is taken as the zero
of energy

The last problem demonstrates that the harmonic approximation, which involves working with the Taylor polynomial of degree 2, provides a totally unrealistic description of the energy when the nuclei make large excursions from R_e. There are two problems here: firstly, we do not possess sufficient information about the interval of convergence for $\mathcal{E}(R)$; secondly, the physical process of bond dissociation cannot be described within this model, as the energy should tend to the value of the bond dissociation energy, D_e, as $R \to \infty$ ▷. In practice, therefore, we need to consider not only extra terms in the Taylor series, but also its interval of convergence to ensure that we can safely use the series for large values of R. By following this procedure we are then able to introduce some element of anharmonicity which is a necessary prerequisite in order describe bond dissociation correctly.

It is common practice to work with a semi-empirically based function ▷ in a so-called Morse form to describe the variation of vibrational energy with R in the case of a diatomic molecule:

$$\mathcal{E}(R) = \mathcal{E}(\infty)\{1 - e^{-a(R-R_e)}\}^2,$$

where $E(R_e)$ is again taken as energy zero so that $\mathcal{E}(\infty) = D_e$.

After multiplying out the bracket on the right-hand side of the expression defining $\mathcal{E}(R)$, a constant and the sum of two exponential functions is obtained; thus, since the Maclaurin series for each of the latter converge for all physically reasonable R, the Maclaurin series for the Morse function itself will also converge for all physically reasonable R. Despite this observation that the Maclaurin series converges for all values of R, a problem arises for large R, where the number of terms needed to obtain good accuracy can be very large indeed. It should also be noted that the range of R values for which a Taylor polynomial of low degree (two or three) provides a good fit to the Morse function is necessarily restricted to a small interval about R_e.

▷

See, for example, Atkins, Section 16.10 or Alberty and Silbey, Section 14.6.

▷

That is, a function which contains experimental data.

▷

Chain and product rules.

Problem 7.12

(a) Use the rules of differentiation ▷ to derive the first four derivatives of the Morse function $\mathcal{E}(R)$ at $R = R_e$. From the pattern observed, write down the rth derivative.

(b) Write down the Taylor series for $\mathcal{E}(R)$ about the point R_e.

(c) Given that the Taylor polynomial of degree two (harmonic approximation) is usually expressed in the form $\dfrac{k}{2}(R - R_e)^2$, where k is the (harmonic) force constant (the stiffness of the bond to stretching), show that

$$a = \sqrt{\frac{k}{2D_e}}.$$

(d) Use the relation between k, the bond stretching frequency, v, and the nuclear reduced mass, $\triangleright$ μ, in the form $v = \dfrac{1}{2\pi}\sqrt{\dfrac{k}{\mu}}$ to express a in terms of v, D_e, and μ.

(e) Given that the conversions of frequency and energy to wavenumber (m^{-1}) units are achieved by the substitutions $v = c\tilde{v}$ and $D_e = \tilde{D}_e hc$, respectively, demonstrate that $\triangleright$

$$a = \tilde{v}\left(\frac{2\pi^2 c\mu}{h\tilde{D}_e}\right)^{\frac{1}{2}}.$$

(f) For $^1\text{H}^{35}\text{Cl}$, where a has the value of $1.8672 \times 10^{10}\,\text{m}^{-1}$, and $R_e = 0.1274 \times 10^{-9}$ m, plot values of $\mathcal{E}(R)/\mathcal{E}(\infty) = \left(1 - e^{-a(R_e\delta)}\right)^2$ against $aR_e\delta$, where $R - R_e = R_e\delta$ for $aR_e\delta = -1$ to $+1$ in steps of 0.2.

$\triangleright$
$\mu = \frac{M_1 M_2}{M_1 + M_2}$, where M_1, M_2 are the two nuclear masses.

$\triangleright$
Notice that Atkins and Alberty and Silbey use the same symbol ω to designate frequency in different units; also the latter authors implicitly assume that D_e is also in wavenumber units.

Limits revisited

The Maclaurin expansion for the exponential function can be used in several situations where the limiting behaviour of an expression containing exponential functions is not obvious. Two examples will be considered here: first $\lim\limits_{x\to\infty} x^n e^{-x}$.

The problem with this limit is that it is not immediately obvious whether x^n increases in value faster than e^{-x} decreases in value, or the reverse is true. If the expression is written as

$$\frac{x^n}{e^x} = \frac{x^n}{1 + x + x^2/2! + \cdots + x^n/n! + x^{n+1}/(n+1)! + \cdots}$$

$$= \frac{1}{1/x^n + 1/x^{n-1} + 1/(2x^{n-2}) + \cdots + 1/n! + x/(n+1)! + \cdots},$$

$\triangleright$
The first n terms in the denominator yield $1/n!$ and the remaining terms increase without limit as $x \to \infty$.

then it is seen immediately that its value tends to zero as x becomes indefinitely large $\triangleright$.

> **Problem 7.13**
>
> Use the Einstein expression for the specific heat of a solid (Problem 3.6(b)) to show that $C_v \rightarrow 0$ in the limit as $T \rightarrow \infty$.

The second example involves evaluating the expression

$$\frac{e^x + e^{-x}}{e^x - e^{-x}} - \frac{1}{x}$$

$\triangleright$
Barrow, p. 617.

for small x. This expression arises in treating orientational contributions to electrical and magnetic properties $\triangleright$. If the functions e^x and e^{-x} are approximated by their Taylor polynomials of degree 3, then the expression becomes

$$\frac{(1 + x + x^2/2 + x^3/6) + (1 - x + x^2/2 - x^3/6)}{(1 + x + x^2/2 + x^3/6) - (1 - x + x^2/2 - x^3/6)} - \frac{1}{x}$$

$$= \frac{2 + x^2}{2x + x^3/3} - \frac{1}{x} = \frac{x^3 - x^3/3}{x(2x + x^3/3)},$$

where the last step follows by forming the common denominator. Thus, for small x, where only the $2x^2$ term in the denominator need be considered, the expression reduces to $x/3$ which, in the limit as $x \rightarrow 0$, yields zero, a result which cannot be obtained from the original expression by substituting $x = 0$.

Summary: The techniques of power series expansions have provided a means by which functions can be represented, and thus approximated, for values of the independent variable within the interval of convergence for the series. Power series are used widely in a chemical context. They are also used in a return visit to complex numbers in the next chapter.

8 Complex numbers revisited

Objectives

This chapter

- introduces the properties and representations of complex functions in Cartesian and polar forms
- defines fractional powers of complex numbers
- illustrates the use of complex quantities in chemistry

$\triangleright$
That is,
$P(x) = f(x) + ig(x)$,
where $f(x)$ and $g(x)$ are real functions.
$\triangleright$
Duffy, Chapter 3.

This second excursion into the topic of complex numbers provides a suitable preparation for further applications within a chemical context – especially matrix algebra. The imaginary number i plays an important role in many areas of chemistry, primarily because properties associated with wave propagation or behaviour are usually described in terms of functions, $P(x)$, which have a real and an imaginary part $\triangleright$. In instances where P is a function of two or more variables, it can still be written in complex form with real and imaginary parts.

Examples of properties based on the use of complex functions are numerous. For example, in measuring the consequences of the interaction of electromagnetic radiation with incompletely transparent inorganic materials, the refractive index is found to have a real and an imaginary part $\triangleright$, the latter being associated with the extent of energy absorption. Again, in interpreting X-ray data using the structure factor, F (see Problem 8.7), allowance has to be made for its complex form.

In nearly all the situations we meet, the observable property is interpreted in the model used for calculating (or rationalizing) experimental data, in terms of the product of P with its complex conjugate, P^*. Such products may be used without further manipulation (as in the previously cited X-ray example), or within an integral which may determine, for example, a transition probability, a dipole moment or an electron density function.

At the moment, we are not in a position to discuss these chemically relevant examples. It is first necessary to extend the discussion of complex numbers given in Chapter 1, in order to develop some useful tools.

8.1   More manipulations with complex numbers

Complex numbers were introduced in Chapter 1 through examining the solution sets of quadratic equations. In another context, if we are concerned with finding the values of real x for which a function such as

$$f(x) = \left(\frac{2+x}{1-x}\right)^{\frac{1}{2}} \tag{8.1}$$

is defined, we require $(2+x)/(1-x) \geq 0$. However, if we want to extend the domain of $f(x)$ to include values of x for which the above inequality is false, then imaginary values of $f(x)$ are obtained. This is seen, for example, in the evaluation of $f(-3)$, when $i/2$ is obtained.

Since complex numbers can arise in extending the domain of real functions, we have to develop ways of manipulating and visualizing functions represented by sets of ordered pairs of real (x) and complex $(f(x))$ numbers; other functions may be defined in which the domain is also the set of complex numbers $\triangleright$.

▷
Usually written as $\mathbb{C}$.

As seen in Chapter 1, addition, subtraction and multiplication of complex numbers in the form $a + ib$ presents no difficulties, as we just apply ordinary rules of algebra, remembering that $i^2 = -1$ (since $i = \sqrt{-1}$).

▷
Hint: write $z_3 = a + ib$
and equate real and
imaginary parts of
$z_1 z_3 = 1$.

> **Problem 8.1**
>
> If $z_1 = 1 - i\sqrt{2}$, and $z_2 = 2 + i\sqrt{2}$, determine $z_1 z_2$, $z_1 + 2z_2$, $z_1 - z_2$, and z_3, such that $z_1 z_3 = 1$ $\triangleright$.

To carry out division of the complex numbers $z_1 = a + ib$, $z_2 = c + id$, the key step involves making the denominator real by multiplying numerator and denominator by the complex conjugate of the denominator, z_2^*:

▷
Remember that
$z_2 z_2^* = (c + id)(c - id)$
$= c^2 + d^2$ is real.

$$\frac{z_1}{z_2} = \frac{z_1 z_2^*}{z_2 z_2^*} \triangleright.$$

Worked example

8.1   Find the quotient of the two complex numbers z_1, z_2 given in Problem 8.1.

Solution

$$\frac{z_1}{z_2} = \frac{(1 - i\sqrt{2})}{(2 + i\sqrt{2})} = \frac{(1 - i\sqrt{2})}{(2 + i\sqrt{2})} \frac{(2 - i\sqrt{2})}{(2 - i\sqrt{2})} \Rightarrow \frac{(1 - i\sqrt{2})(2 - i\sqrt{2})}{6}$$

$$= -\frac{3\sqrt{2}i}{6} = -\frac{i}{\sqrt{2}}.$$

Problem 8.2

Express the following expressions in the form $x + iy$:

(a) $(2 - 3i) + (3 + 4i)/i$, (b) $\dfrac{(1-i)}{(1+i)}$, (c) $(1+i)^2$,

(d) $\dfrac{(1-i)}{(1+i)(2-i)}$, (e) $\dfrac{1}{\cos\beta + i\sin\beta}$.

8.2   Cartesian and polar representations of complex numbers

Complex numbers in the form $z = x + iy$ can be represented using an *Argand diagram*, in which the real and imaginary parts (x and y, respectively) are taken to lie along the x and y axes as shown in Figure 8.1. The position of z can be specified in Cartesian form (x, y) or in polar form (r, θ) where the polar coordinates r and θ provide an alternative way of specifying the position of any point in the plane. The advantage of the Cartesian form is that, given

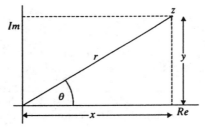

Figure 8.1
Argand diagram for displaying the complex number $z = x + iy$, with real (*Re*) and imaginary (*Im*) parts x and y, respectively

values for x, y, it is very easy with graph paper, for example, to mark off horizontal and vertical distances of x and y, respectively, to define the position of the point z; in the polar coordinate system, however, it is first necessary to find the values for r and θ before the point can be located. From Figure 8.1, it can be seen that the x and y coordinates of the point representing z are given by

$$x = r\cos\theta, \ y = r\sin\theta,$$

from the properties of right-angled triangles. Thus, an arbitrary complex number, z, may be written in the form

$$z = x + iy = r(\cos\theta + i\sin\theta),$$

where r and θ are termed the *modulus* and *argument* of z.

Since

$$zz^* = r^2(\cos\theta + i\sin\theta)(\cos\theta - i\sin\theta) = x^2 + y^2 = r^2,$$

we see that the modulus of z, written $|z|$, is obtained directly from the values of x and y:

$$r = |z| = \sqrt{x^2 + y^2}.$$

Thus, as x, y and r are now known, θ, or $\arg z$, can be obtained from either $x = r \cos \theta$ or $y = r \sin \theta$; however, using either of these equations by itself is not sufficient to determine θ, because $\cos \theta$ is positive, for example, in the first *and* and fourth quadrants. Similarly $\sin \theta$ is positive in the first and second quadrants. It is therefore necessary to fix the quadrant by examining the *sign* of both $\cos \theta$ and $\sin \theta$; also, by convention, θ is specified in the range $-\pi < \theta \leq \pi$ rather than $0 < \theta \leq 2\pi$ (positive and negative angles are measured anticlockwise and clockwise, respectively, from the x-axis).

Worked example

8.2   Find the modulus and argument of the complex number $1 - i$.

Solution $z = 1 - i \Rightarrow z^* = 1 + i;$ hence $zz^* = (1 - i)(1 + i) = 1 - i^2 = 2 = r^2$
$\Rightarrow r = \sqrt{2}.$

Notice that, when taking the square root of 2, the two values $\pm\sqrt{2}$ are obtained; however, r cannot be negative, as it is the distance of the point z from the origin. Hence, on rewriting z in polar form

$$z = r(\cos \theta + i \sin \theta) = \sqrt{2}\left(\frac{1}{\sqrt{2}} - \frac{i}{\sqrt{2}}\right),$$

it follows that

$$\cos \theta = \frac{1}{\sqrt{2}} \text{ and } \sin \theta = -\frac{1}{\sqrt{2}}.$$

θ is therefore located in the fourth quadrant, and equal to $-\pi/4$.

> **Problem 8.3**
>
> Place the solutions to Problems 8.1 and 8.2(a)–(c) on an Argand diagram and also give the values for r, θ associated with each complex number.

8.3   Euler's theorem

The Maclaurin series for the exponential, cosine and sine functions, derived in Chapter 7, are as follows (using θ rather than x for the variable):

$$e^\theta = 1 + \theta + \frac{\theta^2}{2!} + \frac{\theta^3}{3!} + \cdots + \frac{\theta^n}{n!} + \cdots$$

$$\cos\theta = 1 - \frac{\theta^2}{2!} + \frac{\theta^4}{4!} + \cdots + (-1)^n \frac{\theta^{2n}}{(2n)!} + \cdots$$

$$\sin\theta = \theta - \frac{\theta^3}{3!} + \frac{\theta^5}{5!} + \cdots + (-1)^{n-1} \frac{\theta^{2n-1}}{(2n-1)!} + \cdots .$$

If we now rewrite the cosine and sine series, by substituting i^2 for -1, and add the former to i times the latter, we obtain

$$\left\{ 1 + \frac{(i\theta)^2}{2!} + \frac{(i\theta)^4}{4!} + \cdots + \frac{(i\theta)^{2n}}{(2n)!} + \cdots \right\}$$

$$+ \left\{ i\theta + \frac{(i\theta)^3}{3!} + \frac{(i\theta)^5}{5!} + \cdots + \frac{(i\theta)^{2n-1}}{(2n-1)!} + \cdots \right\}$$

which, after rearrangement, becomes

$$\left\{ 1 + i\theta + \frac{(i\theta)^2}{2!} + \frac{(i\theta)^3}{3!} + \frac{(i\theta)^4}{4!} + \frac{(i\theta)^5}{5!} + \cdots \right.$$

$$\left. + \frac{(i\theta)^{2n-1}}{(2n-1)!} + \frac{(i\theta)^{2n}}{(2n)!} + \cdots \right\} . \tag{8.2}$$

This series is evidently the Maclaurin expansion for $e^{i\theta}$. Hence,

$$e^{i\theta} = \cos\theta + i\sin\theta , \tag{8.3}$$

a result which is known as *Euler's theorem*. The use of Equation (8.3) enables the polar form of a complex number to be rewritten in the compact form

$$z = re^{i\theta} \tag{8.4}$$

with complex conjugate $z^* = re^{-i\theta}$. These forms for z and z^* lead to a simplification in the manipulation of complex numbers, as we shall see shortly.

Problem 8.4

For complex numbers of unit modulus, where $z = e^{i\theta}$ and $z^* = e^{-i\theta}$, show that

(a) $\cos\theta = \frac{1}{2}(e^{i\theta} + e^{-i\theta})$ and $\sin\theta = \frac{1}{2i}(e^{i\theta} - e^{-i\theta})$

(b) $\cosh\theta = \cos i\theta$ and $\sinh\theta = -i\sin i\theta$

(c) $\cos^2\theta = \frac{1}{2}(\cos 2\theta + 1).$

If the number system is extended to include complex numbers, the trigonometrical functions may be computed for imaginary values of the independent variable; furthermore, the values of the functions are no longer restricted to lie in the interval $[-1, 1]$. This situation is examined in the next problem.

Problem 8.5

Find the value of x for which $\cos ix = 5$ by solving the equation $\cosh x = 5$ in the following sequence of steps:

(a) use the definition of $\cosh x$ to show that this implies
$e^{2x} - 10e^x + 1 = 0$;
(b) let $e^x = w$, and solve the resulting quadratic equation for w;
(c) for each value of w, find the corresponding value of x.

8.4   Powers of complex numbers: the de Moivre theorem

As noted earlier, the Euler result provides us with a useful and compact way of writing a complex number. One advantage in using the polar rather than the Cartesian form of z is that, for positive integer n,

$$z^n = [r^n(\cos\theta + i\sin\theta)^n] = [r^n(e^{i\theta})^n] = [r^n(e^{in\theta})]$$
$$= r^n(\cos n\theta + i\sin n\theta), \tag{8.5}$$

where the last step follows from the Euler result, Equation (8.3). It therefore follows that

$$(\cos\theta + i\sin\theta)^n = \cos n\theta + i\sin n\theta, \tag{8.6}$$

which is a statement of *deMoivre's theorem*. Furthermore, examination of the last equality in Equation (8.5) reveals that the modulus and argument of z^n are r^n and $n\theta$, respectively.

Extension of the de Moivre result to negative and rational powers

The extension of Equation (8.6) to include negative integers is made by substituting $n = -m$, remembering that $\cos(-m\theta) = \cos(m\theta)$ and $\sin(-m\theta) = -\sin(m\theta)$ ▷, to obtain

$$(\cos\theta + i\sin\theta)^{-m} = (\cos m\theta - i\sin m\theta).$$

▷
See Appendix 2,
equations (A2.10)
(A2.18).

▷
Note that the qth root has the required property that its qth power yields $e^{i\theta}$.

Likewise, for a rational power of the form p/q, if we define a qth root to have the property that ▷

$$(e^{i\theta})^{\frac{1}{q}} = e^{i\frac{1}{q}\theta} = \cos\left(\frac{1}{q}\theta\right) + i\sin\left(\frac{1}{q}\theta\right),$$

then

$$(e^{i\theta})^{\frac{p}{q}} = \cos\left(\frac{p}{q}\theta\right) + i\sin\left(\frac{p}{q}\theta\right).$$

Problem 8.6

Use the Euler and the de Moivre theorems to write the following expressions in the form $a + ib$:

(a) $\left(\cos\frac{\pi}{3} + i\sin\frac{\pi}{3}\right)\left(\cos\frac{\pi}{6} + i\sin\frac{\pi}{6}\right)$, (b) $\left(\cos\frac{2\pi}{3} + i\sin\frac{2\pi}{3}\right)^3$
(c) $(1 - i\sqrt{3})^{10}$ (d) $(1 - i\sqrt{3})^{-6}$.

Hint: In parts (c) and (d), first express $1 - i\sqrt{3}$ in polar form.

Problem 8.7

▷
Alberty and Silbey, p. 800.

The intensity of the scattered beam of X-rays from the (hkl) plane of a crystal is proportional to FF^*, where F, the structure factor, is given by

$$F(hkl) = \sum_{j}^{\text{cell}} f_j e^{2\pi i[hx_j + ky_j + lz_j]} \tag{8.7}$$

and the summation runs over the appropriate number of atoms in the unit cell with (fractional) coordinates (x_j, y_j, z_j) and scattering factor f_j.

▷
That is, a complex function.

(a) By using the Euler result for each exponential factor in the summation in Equation (8.7), show that $F(h\,k\,l)$ may be written in the form $F_1 + iF_2$ ▷, and give the forms for both F_1 and F_2.

(b) For the body-centred cubic structure, the unit cell has two A atoms per unit cell, with one at $(0,0,0)$ and the other at $(\frac{1}{2},\frac{1}{2},\frac{1}{2})$. Thus

$$F(hkl) = f_A[1 + e^{\pi i(h+k+l)}].$$

1. Show that $F(h\,k\,l) = 0$ or $2f_A$ depending whether $h + k + l$ is either an odd or even integer, respectively.

2. Determine which of the following $(h\,k\,l)$ reflections are missing from the X-ray diffraction pattern of metallic sodium: $(300), (200), (111), (222)$

3. In the closely related caesium chloride structure, where the A- and B-type atoms are at $(0,0,0)$ and $(\frac{1}{2},\frac{1}{2},\frac{1}{2})$, respectively,

explain why the reflections that would be missing for a body-centred cubic structure now appear with low intensity.

(c) For the face-centred cubic structure, with four atoms per unit cell, the structure is generated by atoms situated at $(0,0,0)$, $(0,\frac{1}{2},\frac{1}{2})$, $(\frac{1}{2},0,\frac{1}{2})$, and $(\frac{1}{2},\frac{1}{2},0)$.

1. Under what conditions does $F(hkl)$ vanish?
2. The NaCl structure, which contains eight atoms in the unit cell, has four sodium species at the above four sites and four chlorine species at $(\frac{1}{2},0,0)$, $(\frac{1}{2},\frac{1}{2},\frac{1}{2})$, $(0,0,\frac{1}{2})$, $(0,\frac{1}{2},0)$. Give the conditions for $F(hkl)$ to vanish, and compare your expectation with the experimental data illustrated in Barrow $\triangleright$.

$\triangleright$

Fig. 16.11.

8.5   Roots of complex numbers

So far in this chapter we have looked at the addition, multiplication, division, and formation of integer and rational powers of complex numbers – the latter through the use of the Euler and de Moivre theorems.

$\triangleright$

The principal root.

In the formation of rational powers of a complex number, we have used only one of the qth roots of z $\triangleright$, since it had the property that its qth power yielded z. In general, however, there are q qth roots, and it is necessary to investigate how *all* such roots of a real or complex numbers are obtained. For example, the cube roots of -27 are three in number, and -3 is the principal (real) root; the other two roots are complex. Even more strangely, perhaps, the three cube roots of unity are 1, the principal (real) root, and two complex roots. Our problem now is to find a way of identifying all the roots.

The first step in finding the qth roots of a complex number (such numbers include real numbers) is to express it in the general polar form

$$z = r\{\cos(\theta + 2\pi n) + i\sin(\theta + 2\pi n)\} = re^{i(\theta + 2\pi n)}, \qquad (8.8)$$

$\triangleright$

See Appendix 2.

where $n = 0, \pm 1, \pm 2, \ldots$ and, until now, only the case where $n = 0$ has been considered. The extension of the argument of z by multiples of 2π apparently yields nothing new, since $\cos(\theta + 2\pi n) = \cos\theta$ and $\sin(\theta + 2\pi n) = \sin\theta$ $\triangleright$. However, the use of the generalized polar form for z is necessary in order to recover all the other roots when the de Moivre theorem is used, as is seen in the two examples that follow.

Worked example

8.3   (a) Determine the cube roots of unity.

Solution   Taking z in the general form $1\{\cos{(0+2\pi n)}+i\sin{(0+2\pi n)}\}$, de Moivre's theorem yields

$$z^{\frac{1}{3}} = \{\cos{(2\pi n/3)}+i\sin{(2\pi n/3)}\} = e^{2\pi ni/3},$$

where the three cube roots of z are $1, -\frac{1}{2}(1 \pm i\sqrt{3})$, associated with values of n equal to $0, \pm 1 \triangleright$. The consideration of the other values of $n = \pm 2, \pm 3, \ldots$ is not necessary, as there is only repetition of the roots already found.

▷
It is a simple matter to check that the cube of each of these roots yields unity.

(b) Determine the cube roots of $z = -27$.

Solution   Since $z = -27 = 27(-1) = 27\{\cos{(\pi + 2\pi n)}+i\sin{(\pi + 2\pi n)}\}$, we see that the cube roots of z are found in the form

$$z^{\frac{1}{3}} = 3\{\cos{[\pi(1+2n)/3]}+i\sin{[\pi(1+2n)/3]}\}$$
$$= 3e^{i\pi(1+2n)/3} = 3e^{i\pi/3}\cdot e^{2n\pi i/3}.$$

Taking $n = 0, \pm 1$ generates the three cube roots $-3, \frac{3}{2}(1 \pm i\sqrt{3})$.

In both of these examples, it should be noted that the complex roots come in pairs of complex conjugates.

Problem 8.8

Find

(a) the two roots of $x^2 + 1 = 0$
(b) the argument and modulus of i, and hence determine the two square roots of i

▷
See Example 5.3.

(c) the cube roots of $(1 + i)$
(d) the five roots of $x^5 - 2 = 0 \triangleright$.

Logarithms revisited

Before concluding this chapter, it is interesting to see how the general expression for z can provide a way for defining the logarithms of complex numbers. From the general form of a complex number given in Equation (8.8), and using the Euler result, we see that

$$z = re^{i(\theta+2\pi n)}.$$

▷
ln ab = ln a + ln b.

Thus, from the properties of logarithms $\triangleright$, it follows that $\ln z = \ln r + i(\theta + 2\pi n)$, where, for $n = 0$, the principal value of the logarithm is defined.

Problem 8.9

Give the principal value of ln z for the following complex numbers:

(a) $z = -1$, (b) $z = 1 + i$, (c) $z = i$.

Notice that when the number system is extended from the set $\mathbb{R}$ to $\mathbb{C}$, it becomes possible to define the logarithm of a negative number.

Summary: This second excursion into the topic of complex numbers provides a suitable preparation for further applications within a chemical context. The techniques for extending the real number system to include complex numbers of the form $x + iy$ have been seen to provide a way of obtaining additional solutions of equations that are not available in the real number system. In addition, the Euler and de Moivre results enable powers and roots of complex numbers in polar form to be obtained with very little effort.

The next chapter is concerned with differential equations and their solution.

9 The solution of simple differential equations – the nuts and bolts of kinetics

Objectives	This chapter demonstrates how selected tools of integral calculus are deployed in determing the solution of equations depending upon rates of change (derivatives): in particular

- the basic differential equations associated with first-order and sequential first-order chemical processes

- second-order differential equations and their use, extending the techniques developed for solving first-order equations

In chemistry, the study of changing quantities with respect to time or position coordinates forms the basis of a considerable amount of chemical exploration related to structure and mechanism: for example

- in determining the geometrical structure of a molecule in the crystalline state, the effects of thermal motion, as seen in the vibrational and rotational motions of the molecule, requires the use of some kind of time-averaging procedure in model calculations in order to process experimental observations to yield useful information about bond lengths and interbond angles;

- in studies of chemical reactions, kinetic measurements are made in order to check their consistency with postulated mechanisms (to complicate the issue, it can happen that two mechanisms yield the same kinetic observation!);

- understanding how electromagnetic radiation interacts with molecular species, and causes electronic, vibrational or rotational excitation, involves the consideration of the periodic disturbance of the molecular system by the associated electric and magnetic fields.

In all these, and many other, situations we are dealing with phenomena that involve quantities changing either at a characteristic rate in some instances, or in a less well-defined way. The problem lies in the fact that we essentially *observe* a derivative function, in the form of a rate of change of a property with respect to another variable (time, say). We therefore need a branch of mathematics which deals with relations between properties and their rates of change with respect to one or more independent variables (time, spatial coordinates, concentration, etc.) – very often in the presence of an external disturbance. Relations of this kind constitute ordinary differential equations when there is only one independent variable; when there are two or more variables (see Chapters 1 and 10), we are led into the field of partial differential equations ▷. The latter types of equation are fundamental to the development of the quantum mechanical treatments of electronic motion or the vibrations of the nuclei in a molecular species, for example, but we do not focus on these kind of equations here: principally because the inherent symmetry of the system under study usually ensures that the partial differential equation for determining the respective solutions factorizes into a set of differential equations depending upon only one independent variable. However, in treating the electronic motion in an atom, and in many other situations, the differential equations obtained are not 'ordinary', and their solution is obtained by using a power series representation for the solution (a short excursion into this technique is described in the concluding section of this chapter).

Depending upon the complexity of the chemical process under examination, the range of ordinary differential equations may extend from a single equation, which may admit of a straightforward solution, to a complex set of coupled equations involving the rates of change of several variables. However, in order to develop an understanding of both the origin and the solution of differential equations, without being exhaustive, we restrict the material in this chapter to the solution of first-order and selected second-order differential equations of chemical interest. Details of methods for solving the more specialized forms of differential equation, arising in a chemical context, are available in the mathematics literature ▷.

▷
Differential equations which contain partial derivatives.

▷
See, for example, Lapwood.

9.1 First-order differential equations

The idea of a first-order differential equation is, in fact, embodied in the study of the indefinite integral of $f(x)$:

$$\int f(x)dx.$$

▷
As seen in Chapter 6,
integration may be
viewed as
antidifferentiation.

The indefinite integral is determined by finding an antiderivative function $F(x)$ which, upon differentiation, yields $f(x)$ ▷:

$$\frac{dF(x)}{dx} = f(x). \tag{9.1}$$

This is a *first-order* differential equation, since only the first derivative of $F(x)$ appears in the equation. In more conventional notation, if we write $y = F(x)$ (forgetting about the constant of integration for the moment), then the differential equation (9.1) becomes:

$$\frac{dy}{dx} = f(x). \tag{9.2}$$

As we have seen in Chapter 6, all functions in the form $y = F(x) + C$ (where C is an arbitrary constant), satisfy Equation (9.2), and different members of this family of solutions correspond to different choices of the constant of integration, C.

An example of a first-order differential equation in a chemical context is provided by a reaction A $\rightarrow$ B in which the rate is proportional to the concentration of A at time t, that is,

$$-\frac{da}{dt} = ka. \tag{9.3}$$

The key point to notice is that the solution of a *first-order* equation involves *one* step of integration, and hence yields one constant of integration. Not surprisingly, the solution of an nth-order differential equation involves n steps of integration and yields n constants of integration. In chemical situations, the constants of integration are usually determined by constraints on the solution ▷. For the example above, if α is the initial concentration of A, then this boundary condition is sufficient to determine the constant of integration, C, in the solution of Equation (9.3) (see Problem 9.2). Thus out of the family of possible solutions only one solution is acceptable – and this is the one satisfying the boundary condition.

▷
Usually termed
boundary conditions.

The material described in this chapter is concerned first with the solution of some first-order differential equations, and then the methodology is extended to the solution of selected important second-order differential equations.

9.2   Separation of variables for first-order differential equations

Suppose we have a first-order differential equation which can be rearranged to take the form

$$\frac{dy}{dx} = g(y)f(x),$$

containing the first derivative of y and a product of functions of x and y (in Equation (9.2) above, $g(y) = 1$). This equation may be rewritten as

$$\frac{1}{g(y)} \cdot \frac{dy}{dx} = f(x)$$

and integration with respect to x is immediate:

$$\int \frac{1}{g(y)} \cdot \frac{dy}{dx} dx = \int f(x) \, dx + C;$$

that is

$$\int \frac{1}{g(y)} dy = \int f(x) \, dx + C, \tag{9.4}$$

See Chapter 5 for the definition of the differential, *dy*.

since ▷

$$dy = \frac{dy}{dx} dx.$$

There are thus two indefinite integrals to evaluate, and it may be envisaged that the two integration constants are combined in the one constant, C; it is sufficient, therefore, to find the two antiderivative functions, and append C as indicated above in Equation (9.4).

Worked example

9.1   Find the solution of the differential equation

$$\frac{dy}{dx} = 3x^2 y.$$

Solution   This equation is of the required form, as the right-hand side may be written as $g(y)f(x)$, where $g(y) = y$ and $f(x) = 3x^2$.
Thus

$$\int \frac{1}{y} dy = \int 3x^2 dx + C$$
$$\Rightarrow \ln y = x^3 + C$$
$$\Rightarrow y = e^{(x^3 + C)} = e^C . e^{x^3}$$
$$= A e^{x^3}$$

where the constant of integration, C, is written in the form $C = \ln A$, where A is a constant.

Problem 9.1

Solve the following first-order differential equations, using the boundary conditions as specified.

(a) $\dfrac{dy}{dx} = y^2 e^x$, where $y = 1$ when $x = 0$.

(b) $\tan x \dfrac{dy}{dx} = y$, where $y = 2$ when $x = \pi/6$.

(c) $\sin y \dfrac{dy}{dx} + x = 0$, where $y = \pi/3$ when $x = 1$.

(d) Express $\dfrac{1}{y(y+1)}$ in partial fractions, and hence solve $\dfrac{x\,dy}{y\,dx} = y + 1$, given that $y = 1$ when $x = 1$. Give your answer in the form $y = F(x)$.

Problem 9.2

Consider the first-order rate process $A \xrightarrow{k} B$, with rate constant k. If α is the initial concentration of A, and a is the concentration of A at time t, then (see Equation 9.3).

$$-\frac{da}{dt} = ka .$$

(a) Solve this equation to determine how a varies with t.
(b) Use the given boundary condition ($a = \alpha$ at time $t = 0$) to find the value of the constant of integration.
(c) If the concentration of species B at time t is x ▷, and the concentration of A is $a = \alpha - x$, show that

$$x = \alpha(1 - e^{-kt}) . \tag{9.5}$$

(d) Rewrite Equation (9.5) to obtain an expression for e^{-kt}; take logarithms, and demonstrate that

$$k = \frac{1}{t} \ln\left(\frac{\alpha}{\alpha - x}\right) .$$

▷
This is the same, of course, as the amount of A reacted.

Problem 9.3

Consider the second-order reaction:

$$A + B \xrightarrow{k} C$$

where α and β are the initial concentrations of A and B, respectively. If after time t, x mol of A have reacted with x mol of B, then the concentrations of A and B are $(\alpha - x)$ and $(\beta - x)$, respectively. Hence, the rate of loss of A at time t is given by

$$-\frac{d(\alpha - x)}{dt} = \frac{dx}{dt} = k(\alpha - x)(\beta - x).$$

(a) Show that $k = \dfrac{1}{t(\alpha - \beta)} \cdot \ln\left[\dfrac{\beta(\alpha - x)}{\alpha(\beta - x)}\right]$, using the boundary condition $x = 0$ at $t = 0$.

(b) For the situation when B is present in excess $(\beta - x \approx \beta, \alpha - \beta \approx -\beta)$, show that the reaction becomes kinetically of the first-order. Give the value of the effective first-order rate constant.

Problem 9.4

The rate equation for the reaction $2NO_2 \xrightarrow{k} N_2O_4$ can be described in terms of the differential equation

$$-\frac{da}{dt} = ka^2,$$

where a is the concentration of NO_2 at time t. The initial concentration of NO_2 is α.

(a) Use the separation of variables method to show that $a = \dfrac{1}{kt + 1/\alpha}$.

(b) Sketch the curve showing how a varies with t.

(c) Demonstrate that the concentration of NO_2 is half of its initial value at a time given by $1/k\alpha$.

The example and the problems considered so far all fall into the category of equations which may be solved by separating the variables. In practice, not surprisingly, some equations are not presented in this recognizable form; however, a simple substitution ▷ is usually sufficient to cast the given equation in terms of x and y into a new differential equation involving x and the intermediate variable, u, which is amenable to this method of solution.

▷
Sometimes not so obvious!

▷

u is a function of x.

Problem 9.5

(a) Use the substitution $y = xu$ ▷ to transform the differential equation

$$x\frac{dy}{dx} + x + y = 0$$ into the equation

$$-\frac{du}{dx} = \frac{1 + 2u}{x}.$$ (9.6)

(b) Solve Equation (9.6) and show that its solution is

$$u = \frac{A}{x^2} - \frac{1}{2}.$$ (9.7)

(c) Eliminate the intermediate variable from Equation (9.7), and show that the solution of the original differential equation is

$$y = \frac{A}{x} - \frac{x}{2}.$$

(d) Find the solution passing through the point $(1, -1)$.

A surface chemistry example

▷

Hinshelwood and Burk.

As one example of the modelling of observed kinetic data, we consider the early work ▷ on the rate of decomposition of ammonia over a platinum surface. The kinetics of the reaction $NH_3 \rightarrow \frac{1}{2}N_2 + \frac{3}{2}H_2$ are modelled by the equation

$$-\frac{d}{dt}p_{NH_3} = k\left(\frac{p_{NH_3}}{p_{H_2}}\right)$$ (9.8)

where p_{NH_3} and p_{H_2} are the partial pressures of NH_3 and H_2, respectively. The aim of the next problem is to show first how this equation can be written in terms of the initial pressure of pure ammonia, p_0, and the total pressure, p, at time t, and then to demonstrate how the transformed equation may be solved.

If there are n mol of NH_3 initially, and n_1, n_2 and n_3 mol of NH_3, N_2 and H_2 at time t, then the initial pressure and partial pressures at time t are given by

$$p_0 = n\frac{RT}{V}, \quad p_{NH_3} = n_1\frac{RT}{V}, \quad p_{H_2} = n_3\frac{RT}{V}, \quad p_{N_2} = n_2\frac{RT}{V},$$

respectively, where V is the volume containing the gases.

To proceed towards the solution of Equation (9.8) we need to reduce the number of variables to two by expressing the three partial pressures in terms of

▷

Pressure and time.

p and p_0; but, as p_0 is a constant, there are effectively just two variables ▷ and the transformed differential equation is then capable of solution.

Assuming that, at a given time, x mol of NH_3 have decomposed, the number of moles of each species is as follows: $n_1 = n - x$, $n_2 = \dfrac{x}{2}$, $n_3 = \dfrac{3x}{2}$.

The total number of moles of species is $n_1 + n_2 + n_3 = n + x$, and hence the pressure is

$$p = (n + x)\frac{RT}{V} = p_0 + x\frac{RT}{V} \Rightarrow x\frac{RT}{V} = p - p_0 .$$

It follows that

$$p_{H_2} = \frac{3x}{2}\frac{RT}{V} = \frac{3}{2}(p - p_0) , \quad p_{N_2} = \frac{x}{2}\frac{RT}{V} = \frac{1}{2}(p - p_0) \text{ and}$$

$$p_{NH_3} = (n - x)\frac{RT}{V} = p_0 - (p - p_0) = 2p_0 - p .$$

▷
Remember that p_0 is a constant.

Thus Equation (9.8) is transformed into ▷

$$-\frac{d(2p_0 - p)}{dt} = \frac{dp}{dt} = \frac{2}{3}k \cdot \frac{2p_0 - p}{p - p_0} . \tag{9.9}$$

Problem 9.6

(a) Show that, by separating the variables and integrating Equation (9.9), we obtain

$$-\int \frac{p_0 - p}{2p_0 - p}\, dp = \frac{2k}{3}\int dt . \tag{9.10}$$

▷
Hint:
$(p_0 - p)/(2p_0 - p) = 1 - p_0/(2p_0 - p)$.

(b) By adding and subtracting p_0 to the numerator of the integrand on the left-hand side, demonstrate that after division ▷ and integration, Equation (9.10) becomes

$$-p - p_0 \ln(2p_0 - p) = \frac{2}{3}kt + C .$$

(c) Use the boundary condition $p = p_0$ at $t = 0$, to determine C.

(d) Show that the solution is

$$\ln\left(\frac{2p_0 - p}{p_0}\right) = 1 - \frac{p}{p_0} - \frac{2k}{3p_0}t .$$

9.3   First-order linear differential equations

A first-order linear differential equation has the form

$$\frac{dy}{dx} + yP(x) = Q(x) , \tag{9.11}$$

where the terms on the left-hand side are of first degree in both y and its first derivative; the special case when $Q(x) = 0$ reduces to the kind of differential equation that may be solved by separating the variables.

A first-order linear differential equation arises in many chemical contexts – the most common one being in the modelling of the kinetics of consecutive first-order reactions.

Consider the sequence of first-order reactions

$$A \xrightarrow{k_1} B \xrightarrow{k_2} C,$$

where a, b, c are the concentrations of A, B, C at time t. The rate of appearance of B involves two terms:

$$\frac{db}{dt} = k_1 a - k_2 b, \tag{9.12}$$

the first one associated with its formation from A, and the second with its loss in the formation of C. But as the first step $A \to B$ is a first-order process, for which $a = \alpha e^{-k_1 t}$ (see the solution to Problem 9.2), Equation (9.12) becomes

$$\frac{db}{dt} + k_2 b = \alpha k_1 e^{-k_1 t}, \tag{9.13}$$

which is in the form of Equation (9.11) (b and t are now the dependent and independent variables, respectively).

$\triangleright$

See Problem 9.7.

Before solving Equation (9.13) to determine how b varies with time $\triangleright$, we recognize that this equation is just a special case of the more general Equation (9.11).

The solution of a first-order linear differential equation

The solution of Equation (9.11) is approached by means of the following well-defined series of steps.

Step 1   Multiply both sides of Equation (9.11) by an as yet undetermined function, defined by $w = R(x)$:

$$w\frac{dy}{dx} + ywP(x) = wQ(x).$$

Step 2   Assuming that $R(x)$ is differentiable, insert the expression $y\dfrac{dw}{dx} - y\dfrac{dw}{dx}$ on the left-hand side so that the derivative of the product yw can be introduced:

$$w\frac{dy}{dx} + y\frac{dw}{dx} - y\frac{dw}{dx} + ywP(x) = wQ(x)$$

$$\Rightarrow \frac{d}{dx}(yw) - y\frac{dw}{dx} + ywP(x) = wQ(x). \tag{9.14}$$

Step 3   Choose $R(x)$ so that

$$\frac{dw}{dx} = wP(x) \tag{9.15}$$

so that the second and third terms on the left-hand side of Equation (9.14) vanish. y is then obtained from the solution of

$$\frac{d}{dx}(yw) = wQ(x) \Rightarrow yw = \int wQ(x)\,dx + C \tag{9.16}$$

after integration. Thus, rewriting w as $R(x)$ yields the solution in the form

$$yR(x) = \int R(x)Q(x)dx + C \Rightarrow y = \frac{1}{R(x)}\int R(x)Q(x)dx + \frac{C}{R(x)}, \tag{9.17}$$

where $R(x)$ remains to be determined.

Step 4   Solve Equation (9.15) to determine $R(x)$, by using the separation of variables technique: ▷

▷

$P(x)$ is known.

$$\int \frac{1}{w}\cdot\frac{dw}{dx}\,dx = \int P(x)dx$$

$$\Rightarrow \int \frac{dw}{w} = \int P(x)dx$$

$$\Rightarrow \ln w = \int P(x)dx$$

i.e., $w = R(x) = e^{\int P(x)dx}$ $\tag{9.18}$

It is not necessary to insert a constant of integration here, as the function $R(x)$ appears on both sides of Equation (9.16); any constant of integration would form a common factor, and thereby cancel out. $R(x)$ is termed the *integrating factor*.

To reiterate: the steps in obtaining the solution of a linear first-order differential equation are

1. Choose $R(x) = e^{\int P(x)\,dx}$.
2. Integrate the product $R(x)Q(x)$.
3. Divide the result by $R(x)$ and add $C/R(x)$ to obtain y.

Worked example

9.2   Determine the solution of the differential equation

$$x\frac{dy}{dx} + x + y = 0$$

given in Problem 9.5.

Solution This equation may be divided through by x to yield $\dfrac{dy}{dx} + y\dfrac{1}{x} = -1$, which enables us to make the following identifications: $P(x) = \dfrac{1}{x}$, and $Q(x) = -1$. Thus the integrating factor is $R(x) = e^{\int (1/x)dx} = e^{\ln x} = x$, and the solution is obtained in the form

$$y = \frac{1}{x}\int -x\,dx + \frac{C}{x} = -\frac{x}{2} + \frac{C}{x},$$

which agrees with the answer given in Problem 9.5.

Sequential first-order reactions revisited

The next problem is concerned with the solution of the differential equation for the series of sequential first-order reactions, given in Equation (9.13). The first point to notice is that the differential equation is posed in terms of the independent and dependent variables t and b, respectively $\triangleright$.

$\triangleright$
If the change in variable names from the conventional x and y causes problems, think in terms of dependent and independent variables!

Problem 9.7

(a) Show that the integrating factor is given by $e^{k_2 t}$.
(b) Use the integrating factor to solve Equation (9.13), and verify that
$$b = \frac{k_1 \alpha}{k_2 - k_1}e^{-k_1 t} + Ce^{-k_2 t}.$$ Given the boundary condition $b = 0$, $t = 0$, determine C, and hence find the expression for b as a function of time, t.

This concludes our brief survey of first-order differential equations. The next section is concerned with a number of relevant second-order differential equations that are useful within a chemical context; however, we make no apology for a brief look at a subject which is of far greater range and scope than we can possibly hope to cover in this introductory text.

9.4   Second-order differential equations

Simple harmonic motion

The equation for simple harmonic motion forms a very important paradigm, and arises in areas of physics, chemistry and engineering that are concerned with modelling chemical or mechanical systems under the influence of an oscillatory or periodic change. Thus, for example, the nuclei of a molecular species undergo vibrations, which may be modelled by a system of springs with differing characteristic stiffness – some of which are loosely coupled (to mimic bond–bond interaction).

▷
In a mechanical example, x and y would correspond to time and spatial displacement, respectively.

The equation itself takes the mathematical form ▷

$$\frac{d^2 y}{dx^2} = -\omega^2 y, \tag{9.19}$$

where ω is a constant. If y is a length (displacement) and x is time then the second derivative of y corresponds to a length divided by the square of the time; comparison with the right-hand side shows immediately that ω is characterized by inverse time, and is therefore identified as a frequency. Equation (9.19) is an example of a *homogeneous* second-order differential equation with constant coefficients, as the equation may be written in the form

$$\frac{d^2 y}{dx^2} + \omega^2 y = 0, \text{ with a zero right-hand side, and a weighted sum of}$$

▷
In this case, second and zeroth-order derivatives.

derivatives with respect to x on the left-hand side ▷.

Since $d^2 y / dx^2$ is the second derivative of y, we can write Equation (9.19) in terms of the $\hat{D}$ operator first introduced in Chapter 4:

$$\frac{d^2 y}{dx^2} \equiv \frac{d}{dx}\left(\frac{dy}{dx}\right) \equiv \frac{d}{dx}\left(\frac{d}{dx}y\right) = \hat{D}^2 y = -\omega^2 y.$$

Hence the differential equation becomes

$$(\hat{D}^2 + \omega^2)y = 0,$$

when expressed in $\hat{D}$ operator form.

The term in brackets on the left-hand side of this equation can be factorized as follows (note that, here, the roots are complex):

$$(\hat{D} + i\omega)(\hat{D} - i\omega)y = 0 \tag{9.20}$$

where i is the imaginary number introduced in Chapter 1 and discussed further in Chapter 8.

If the intermediate variable z is introduced by means of the substitution

$$(\hat{D} - i\omega)y = z \tag{9.21}$$

then Equation (9.20) becomes

$$(\hat{D} + i\omega)z = 0, \tag{9.22}$$

which can, in fact, be rewritten as a simple first-order differential equation whose solution is obtained by separating the variables. Thus,

$$(\hat{D} + i\omega)z = 0 \Rightarrow \frac{dz}{dx} + i\omega z = 0$$

$$\Rightarrow \int \frac{dz}{z} = -i \int \omega \, dx$$

$$\Rightarrow \ln z = -i\omega x + C$$

$$\Rightarrow z = A e^{-i\omega x}.$$

The equation defining z in terms of x is now substituted into Equation (9.21) to obtain

$$\frac{dy}{dx} - i\omega y = Ae^{-i\omega x},$$

which is of the form of a linear first-order differential equation for determining y, with $P(x) = -i\omega$ and $Q(x) = Ae^{-i\omega x}$.

The integrating factor is therefore

$$e^{-\int i\omega dx} = e^{-ix\omega},$$

and the solution follows from Equation (9.17):

$$ye^{-i\omega x} = \int e^{-i\omega x} Ae^{-i\omega x} dx + C$$

$$\Rightarrow ye^{-i\omega x} = \frac{Ae^{-2i\omega x}}{-2i\omega} + C \Rightarrow y = Be^{-i\omega x} + Ce^{i\omega x},$$

▷

Notice that Euler's theorem (Equation (8.3)) can be used to express the solution in the form $y = a \cos(\omega x) + b \sin(\omega x)$.

where B, C are constants ▷. As noted in the opening section of this chapter, two constants of integration are expected in specifying the general solution of a *second-order* differential equation.

Problem 9.8

(a) Find the general solution of the differential equation

$$\frac{d^2 y}{dx^2} - 5\frac{dy}{dx} + 6y = 0, \tag{9.23}$$

by introducing first the $\hat{D}$ operators and then an appropriate intermediate variable, z.

Given the boundary conditions $y = 0, x = 0$ and $dy/dx = 1$, $x = 0$, show that the appropriate solution is $y = e^{3x} - e^{2x}$.

(b) Show that the general solution of the equation $\dfrac{d^2 y}{dx^2} = \lambda^2 y$ is

$$y = A \sinh \lambda x + B \cosh \lambda x.$$

▷

Including, as seen above, y, which can be regarded as the zeroth-order derivative.

▷

See, for example, Lapwood, Section 5.3.

The differential equations considered in the example and problem above are of a type in which constant coefficients multiply each of the derivatives of y ▷. When the coefficients of the three derivatives are polynomial functions of the independent variable, then one route towards the solution involves the introduction of an intermediate variable in order to reduce the equation to a form involving constant coefficients (if this is possible). There are no new principles here, and it is best to consult a specialized text for examples and further information ▷.

Inhomogeneous second-order differential equations

A more important kind of second-order differential equation is provided by an extended form of those in Problem 9.8, or the equation for simple harmonic motion, in which the right-hand side is not zero but a function of the independent variable. In the latter example, the effects of anharmonicity (deviation from harmonic behaviour) in molecular vibrations could be included in this way. This extended form of a second-order differential equation is termed an *inhomogeneous* differential equation.

The $\hat{D}$ operator method, used with the introduction of an intermediate variable, is also applicable to inhomogeneous second-order differential equations with constant coefficients. The procedure is straightforward, and leads to a separation of the original equation into two first-order linear differential equations in the process of obtaining the general solution.

Worked example

9.3   Consider the homogeneous differential equation in Problem 9.8 in which the right-hand side now contains the function $Q(x) = x$ to yield the inhomogeneous differential equation:

$$\frac{d^2y}{dx^2} - 5\frac{dy}{dx} + 6y = x. \tag{9.24}$$

Solution   This equation can be rearranged using the same technique as before:

$$(\hat{D}^2 - 5\hat{D} + 6)y = x$$
$$\Rightarrow (\hat{D} - 3)(\hat{D} - 2)y = x$$
$$\Rightarrow (\hat{D} - 3)z = x,$$

where the substitution

$$(\hat{D} - 2)y = z \tag{9.25}$$

is used.

The first step is to solve the first-order linear equation

$$\frac{dz}{dx} - 3z = x \tag{9.26}$$

to determine z as a function of x. The integrating factor is just e^{-3x}; hence, using Equation (9.17),

$$z = e^{3x} \int x e^{-3x} dx + Be^{3x}$$

$$= e^{3x} \left\{ -\frac{xe^{-3x}}{3} + \frac{1}{3} \int e^{-3x} dx \right\} + Be^{3x} \text{ (using integration by parts)}$$

$$\Rightarrow z = -\frac{x}{3} - \frac{1}{9} + Be^{3x}.$$

Thus, substituting for z in Equation (9.25) yields the second first-order linear differential equation

$$\frac{dy}{dx} - 2y = -\frac{x}{3} - \frac{1}{9} + Be^{3x},$$

where the right-hand side is identified as the function $Q(x)$. The integrating factor is e^{-2x}, and hence the solution follows from another use of Equation (9.17):

$$y = e^{2x} \int e^{-2x}\left(Be^{3x} - \frac{x}{3} - \frac{1}{9}\right) + Ce^{2x}$$

$$= Be^{3x} + \frac{1}{6}x + \frac{1}{12} + \frac{1}{18} + Ce^{2x}$$

$$= Be^{3x} + Ce^{2x} + \frac{1}{6}x + \frac{5}{36}.$$

A comparison of this solution with that obtained in Problem 9.8 for the differential Equation (9.23) demonstrates an important point that arises as a consequence of a theorem in the theory of differential equations: namely, that the solution of the inhomogeneous equation (here with the right-hand side equal to x) contains the solution of the homogeneous equation. The extra terms in the solution of the inhomogeneous equation form what is often referred to as a particular integral. Sometimes, rather than following the procedure as outlined above of introducing an intermediate variable, the homogeneous equation is solved first and then a particular integral is found by inspection.

Problem 9.9

Find the general solution of the differential equation

$$\frac{d^2y}{dx^2} + \frac{dy}{dx} = e^x,$$

and show that the solution satisfying the boundary conditions $y = 0, x = 0$ and $\frac{dy}{dx} = 1, x = 0$ is $y = \sinh x$.

As already indicated earlier in this chapter, there are several important occurrences of first-order differential equations arising out of chemical situations. Apart from the simple first-order processes, such as radioactive decay, and the series of sequential first-order reactions leading to first-order linear equations, there is a plethora of examples where other systems of differential equations arise: for example in chain reactions involving radical intermediates; in second-order rate processes of all kinds, and in more complex processes involving simultaneous equilibria. In all these situations, the methods described here can be developed in an appropriate way to solve such systems of differential equations. However, although some cases are best

▷

Matrices are discussed
in Chapter 13.

handled using matrix-based ▷ methods, there are really no new matters of
principle. We therefore proceed to an alternative route to the solution of
differential equations in which a power series expansion for the solution is
sought.

9.5   Power series solution of differential equations

We have seen in Chapter 7 that power series representations of functions
provide an alternative approach to handling functions – either where
knowledge of the function is required in the close proximity of a given
value of x, or where a closed expression for the function is unknown. Since
there are some important differential equations that have to be solved by power
series methods (especially those with variable coefficients for the derivatives
of various orders), it is useful to recast some of the problems and examples
appearing earlier in this chapter in order to illustrate the general principles of
the approach.

When a solution of a differential equation is sought in the form of a
power series, we have to ask the question, what sort of power series? If the
solution cannot be expressed in terms of a Maclaurin series ▷ then the
approach of seeking a solution in the form

▷

Or, in general, a Taylor
series.

$$y = c_0 + c_1 x + c_2 x^2 + \cdots + c_r x^r + \cdots \tag{9.27}$$

will fail. The power series method works satisfactorily for equations of the
form

$$a \frac{d^2 y}{dx^2} + p(x) \frac{dy}{dx} + q(x) y = 0 \,,$$

but, for certain choices of $p(x)$ or $q(x)$, it may be necessary to search for a
solution with y expanded according to

$$y = x^m (c_0 + c_1 x + c_2 x^2 + \cdots + c_r x^r + \cdots) \,,$$

▷

See, for example,
Stephenson where this,
the Frobenius method, is
discussed.

where the values of m (not necessarily a positive integer) are determined in the
process of solution ▷.

Whichever form of power series is used to determine the solution of the
given differential equation, the problem remains at the end of knowing
whether the series can be expressed more succinctly as a combination of power
series representations of standard functions. Sometimes this is possible, and
the solution can be given in an algebraic form; at other times it is only possible
to work with the infinite series representation of the solution.

Because of the complex nature of this area of the calculus, only a few
examples of solving differential equations by the power series method are
considered here. The important equations associated with the names of
Hermite, Laguerre, Legendre and Bessel produce solutions that are in

widespread use throughout chemistry. The solutions of these classic equations are obtained using only relatively simple extensions of the illustrative example and problems considered here.

A simple example

Consider first, the first-order differential equation in Example 9.1;

$$\frac{dy}{dx} = 3x^2 y. \tag{9.28}$$

As we know from the earlier treatment of this equation, the solution can contain only *one* arbitrary constant; it will be necessary, therefore, to find a way of eliminating all the unknown coefficients, except one, in the power series representation of the function, $y = f(x)$. The procedure for doing this is straightforward, and involves first determining the derivative of Equation (9.27):

$$\frac{dy}{dx} = c_1 + 2c_2 x + 3c_3 x^2 + \cdots + rc_r x^{r-1} + \cdots \tag{9.29}$$

and then equating it with $3x^2$ times y in the form of Equation (9.27):

$$3c_0 x^2 + 3c_1 x^3 + 3c_2 x^4 + \cdots + 3c_{r-3} x^{r-1} + \cdots, \tag{9.30}$$

where it should be noted that the power of x in the general terms specified in Equations (9.30), (9.29) has been taken to involve the $(r-1)$th power ▷. Since the two series are specified to be equal, and the value of x is arbitrary (subject only to the fact that the value must lie in the interval of convergence for the series), we can compare coefficients of different powers of x:

x^0	x^1	x^2	x^3	x^4	x^5	x^{r-1}
$c_1 = 0$	$c_2 = 0$	$3c_3 = 3c_0$	$4c_4 = 3c_1$	$5c_5 = 3c_2$	$6c_6 = 3c_3$	$rc_r = 3c_{r-3}$

The general relations involving the coefficients are $c_1 = 0$, $c_2 = 0$, and $(rc_r - 3c_{r-3}) = 0$, for $r = 3, 4, \ldots$. The iterative use of the latter relation – the so-called *indicial equation* – provides a route to obtaining explicit expressions for c_r ($r \geq 3$); however, before doing this, it is helpful to examine the overall pattern of coefficients.

The table above shows that c_3 is the first non-zero coefficient after c_0; the next one after c_3 is c_6 because c_5 is related to c_2 (which is zero) and c_4 is related to c_1 (which is also zero). The pattern is now clearer: c_r is non-zero only for those values of r that are a multiple of 3; that is $r = 3k$, say. The indicial equation is therefore best rewritten to display only the relation between non-zero coefficients:

$$3kc_{3k} = 3c_{3k-3} \Rightarrow c_{3k} = \frac{1}{k} c_{3(k-1)}, \quad k = 1, 2, 3, \ldots.$$

▷ It does not matter which general power of x is selected – it must involve the same power for each series.

Thus the iterated use of this form of the equation yields

$$c_{3k} = \frac{1}{k} c_{3(k-1)}$$

$$= \frac{1}{k(k-1)} \cdot c_{3(k-2)}$$

$$= \frac{1}{k(k-1)(k-2)} \cdot c_{3(k-3)}$$

$$= \frac{1}{k(k-1)(k-2)\cdots(k-r+1)} \cdot c_{3(k-r)}$$

$$= \frac{1}{k!} \cdot c_0, \text{ on putting } k = r \text{ in the previous expression.}$$

The solution of Equation (9.28) is now found by substituting the values of the coefficients, c_r, in Equation (9.27):

$$y = c_0 \left\{ 1 + \frac{1}{1!}x^3 + \frac{1}{2!}x^6 + \frac{1}{3!}x^9 + \cdots + \frac{1}{r!}x^{3r} + \cdots \right\}. \tag{9.31}$$

Two comments need to be made: first, as expected there is one undetermined constant in the solution; second, it is not immediately apparent that the power series is a representation of one or more standard functions. Fortunately, the latter point is quickly resolved because, after making the substitution $X = x^3$, the series is immediately recognized as that for the function $\exp x^3$; hence the required solution is

$$y = c_0 e^{x^3},$$

as found earlier in Example 9.1.

Problem 9.10

Use the power series, Equation (9.27), to find the solution of the equation $\dfrac{d^2y}{dx^2} = 4y$, using the following steps:

(a) Demonstrate that the second derivative of y can be written as

$$2 \cdot 1\, c_2 + 3 \cdot 2\, c_3 x + 4 \cdot 3\, c_4 x^2 + \cdots + (r+2)(r+1)c_{r+2}x^r + \cdots . \tag{9.32}$$

(b) Compare the coefficients of powers of x in Equation (9.32) with the expansion for $4y$, and show that

$$c_2 = \frac{4}{2 \cdot 1} c_0, \quad c_4 = \frac{4}{4 \cdot 3} c_2, \quad \cdots$$

▷

At this juncture, we see that the solution contains two arbitrary coefficients, as expected.

and

$$c_3 = \frac{4}{3 \cdot 2} c_1, \quad c_5 = \frac{4}{5 \cdot 4} c_3, \quad \dots ,$$

thereby demonstrating that all coefficients with an even or odd suffix can be expressed in terms of either c_0 or c_1, respectively ▷.

(c) Show that, for coefficients with an even suffix, where $r = 2k$,

$$c_{2k+2} = \frac{4}{(2k + 2)(2k + 1)} c_{2k} .$$

(d) Iterate the result in (c) to obtain $c_{2k+2} = \dfrac{4^{k+1}}{(2k + 2)!} c_0$.

(e) Repeat the analysis for the coefficients with an odd suffix, and show that $c_{2k+1} = \dfrac{4^k}{(2k + 1)!} c_1$.

(f) Confirm that the solution of the differential equation takes the form

$$y = c_0 \left(1 + \frac{4}{2!} x^2 + \frac{4^2}{4!} x^4 + \dots + \frac{4^k}{(2k)!} x^{2k} + \dots \right)$$
$$+ c_1 \left(x + \frac{4}{3!} x^3 + \frac{4^2}{5!} x^5 + \dots + \frac{4^k}{(2k + 1)!} x^{2k+1} + \dots \right) .$$

(g) Substitute $x = X/2$, and use the Maclaurin series for e^X and e^{-X} to show that the solution becomes

$$y = \frac{c_0}{2} \left(e^X + e^{-X} \right) + \frac{c_1}{4} \left(e^X - e^{-X} \right) .$$

Write the solution in the form $Ae^{2x} + Be^{-2x}$, giving expressions for A and B.

Problem 9.11

Solve Equation (9.23) by the power series method, and verify that the same answer is obtained as before.

Summary: This chapter concludes our examination of the calculus associated with functions of a single variable. The picture that emerges is complex and extensive, and there are many subtle interrelations between the various facets of the differential and integral calculus in the solution of differential equations. The way the ideas reviewed here impinge on chemistry is illustrated through the choice of suitable examples and relevant problems.

The next chapter is concerned with extending the calculus to functions of two or more variables.

10 Functions of two or more variables – differentiation revisited

Objectives

This chapter

- shows how the calculus of functions of a single variable can be extended to functions of two or more variables
- provides chemically based examples to illustrate the application of the mathematical techniques, with especial emphasis on thermodynamics and quantum chemistry

As we saw in Chapter 2, and subsequent chapters, the study of functions of a single variable extends from the understanding of the idea of function itself to dealing with the tools for manipulating such functions in situations where we are interested in rates of change: thus we are drawn into the need for describing how the operations of differentiation and integration can provide a means for transforming functions into new functions which, in the chemical context, provide essential information and insights into the detailed workings of chemical statics (structure) and dynamics (kinetics). In those situations where the function form is unknown, or is required to be approximated, we also saw how power series representations of functions could provide a valuable tool.

10.1 The representation of functions of two or more variables

▷

As seen in the preview at the end of Chapter 4.

Although we have spent a considerable amount of time discussing the many facets of the properties and manipulation of functions of a single variable, the reality is that, in the chemical context ▷, we have to deal on many occasions with systems whose properties are described in terms of functions of two or more variables. The free energy of gaseous Ar, for example, depends upon (say) the pressure and temperature of the gas. Such a dependence is described in terms of a function of two variables, $G = f(P, T)$, where pressure and temperature are the independent variables and, as for functions of a single variable, f defines the rule or prescription.

The variation of G with T and P can be displayed graphically by drawing a surface, with P, T, G taken as labels for a Cartesian axis system. ▷

> ▷
> See Figure 10.1 for an example of a graphical representation of a function of two variables.

For functions with more than two independent variables, we cannot 'draw' the function in any meaningful way without losing information. However, we could keep all but three of the independent variables constant, and draw surfaces displaying the variation of the dependent variable. It is clear, though, that in many situations, such as that described by the free energy of a mixture of reactive gases A, B, in which a product AB is produced, we have to forego any simple visualization of the function ▷ $G = g(P, T, N_A, N_B, N_{AB})$ in terms of a graphical image. This is why it is so important to think in a more abstract way right from the start and learn how to handle functions of many variables using the tools of mathematics, especially the calculus.

> ▷
> This is a different function from the one used for the free energy of a monatomic gas – hence the different function name.

Coordinate systems for properties depending upon two variables

In Figure 10.1, the property z is displayed using a Cartesian coordinate system, in which the origin corresponds to zero values of all variables. In more general terms, for a property like the electron density in an atom or molecule, depending upon the independent variables x, y, z, the value of the density, $\rho = f(x, y, z)$, requires a fourth dimension in order to display its form.

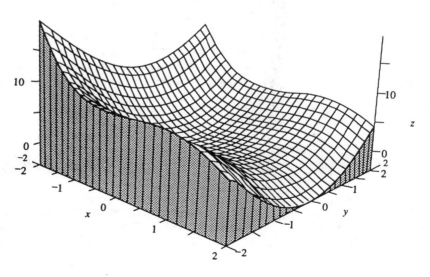

Figure 10.1
Schematic plot of a function of two variables
($z = x^2 + 2y^2 - x^3$)
showing a minimum at (0,0) and a saddle point at $(\frac{2}{3}, 0)$

▷
See Alberty and Silbey,
Figure 12.6.

However, if z is fixed, then values of ρ can be plotted along the axis perpendicular to the xy plane. Thus, for example, if the electron density for the hydrogen molecule ion, in which the nuclei are positioned in the xy plane, is evaluated for points in the plane containing the nuclei ($z = 0$), its value can be plotted at the appropriate height above the plane. The collection of such points forms a surface which shows strong peaks around each nucleus ▷. Different values of z lead to different surfaces.

The origin for the Cartesian coordinate system is taken at one point $(0,0,0)$ where the three axes lie in pairs of intersecting planes. Of the two possible directions for the z-axis, the one is taken for which all three axes form a right-handed system – an arrangement which can be seen by assigning the middle finger, thumb and forefinger to the directions of the x, y, z axes by holding the right hand up with the middle finger pointing forwards, and the thumb directed outwards at right angles to the forefinger. For given constants a, b, c, an arbitrary point is referenced by the three intersecting planes $x = a$, $y = b$, $z = c$, and the length of the line cut off in the three coordinate directions represents the coordinates (a, b, c).

Other coordinate systems, involving the intersection of three different kinds of surface, have different uses – the particular choice depending upon the nature of the problem in question. Another common system is the one provided by *spherical polar coordinates*, where the distance from the origin to a point P is denoted by r (see Figure 10.2). The other two coordinates are θ (the angle between the z-axis and the line passing through O and P, and ϕ (the angle measured between the x-axis and the projection of the line OP onto the xy plane). These two angles are best thought of in geographical terms, where θ is the latitude and ϕ the longitude, when regarding the positions of points on the surface of the earth (r fixed). In the spherical polar coordinate system, the three coordinate surfaces are a sphere (constant r), a cone (constant θ), and a half-plane (constant ϕ).

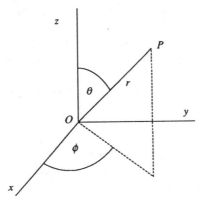

Figure 10.2
The spherical polar
coordinate system

Simple trigonometry yields the following relation between the Cartesian and the spherical polar coordinates of the point P:

$$x = r \sin \theta \cos \phi, \quad y = r \sin \theta \sin \phi, \quad z = r \cos \theta.$$

The orbital wavefunction for describing the motion of the electron in an atom can therefore be described using either system of coordinates; however, the spherical polar system is more in keeping with the spherical symmetry of an atom, and is the preferred choice.

10.2   Differentiation of functions of two or more variables

In this chapter we are concerned with the differentiation and formation of the differential of functions with two or more variables. The search for turning points and the identification of their nature is more problematical than is the case for functions of a single variable: hence, this topic, important though it is, is treated in a pragmatic way. Integration is discussed in the next chapter.

The partial derivative

Before we build on the approach described in Chapter 4 for differentiating a function of two (or more) variables, it is convenient to visualize an area of countryside which is especially undulating, and to focus attention on one particular hill which we have to climb! Imagine that we stand at one point on the surface of the hill and then turn round through 360°: the slope of the terrain will vary according to the direction in which we choose to walk. For example, if we follow a contour there is no change in height and the slope is therefore zero; however, there will be directions where the slope is positive (uphill) or negative (downhill). Let us now return to the free energy analogue of the function shown in Figure 10.1 and consider slicing through the surface with the plane $P = P_0$ ▷. The curve obtained where the surface cuts the plane shows the variation of $f(P_0, T)$ with T. This function can be differentiated in the usual way to find the slope of the tangent at any point (P_0, T): however, as seen in Chapter 4, we indicate that the variable P is held constant by appending it as a subscript to the partial derivative symbol

▷
Analogous to $x = 2$ in Figure 10.1.

$$\left(\frac{\partial G}{\partial T}\right)_P$$

to be more specific ▷. The partial derivative itself is found by a simple extension of the limiting process described in Chapter 4 for a function of a single variable. Thus

▷
Note, again, the use of the 'curly' differentiation symbol to designate the partial derivative.

$$\left(\frac{\partial G}{\partial T}\right)_P = \lim_{h \to 0}\left(\frac{f(P_0, T + h) - f(P_0, T)}{h}\right)$$

gives the slope of the tangent to the surface in the T-direction at the point (P_0, T). Similarly, we may define $(\partial G/\partial P)_T$.

In general, of course, because we deal with variables other than P, T and G, it is helpful to move over to the non-specific notation for the partial derivatives and use variables named x, y and z whilst working through the basic ideas of partial differentiation. Thus, for the function $z = f(x, y)$, the partial derivatives of z are defined by

$$\frac{\partial z}{\partial y} = \lim_{h \to 0} \left(\frac{f(x, y + h) - f(x, y)}{h} \right), \quad \frac{\partial z}{\partial x} = \lim_{h \to 0} \left(\frac{f(x + h, y) - f(x, y)}{h} \right).$$

In determining $\partial z / \partial x$, for example, y is held constant, and the differentiation is carried out with respect to x. The respective value of each partial derivative depends upon the choice for x and y. In general, therefore, the partial derivatives are new functions of x and y.

Worked example

10.1 If $z = f(x, y) = x^2 y + y$, find $\dfrac{\partial z}{\partial x}$ and $\dfrac{\partial z}{\partial y}$ at the points $(1, 2)$ and $(0, 0)$.

Solution Using the method described in Chapter 4, $\partial z / \partial x = 2xy$ and $\partial z / \partial y = x^2 + 1$. Hence, at $(1, 2)$, $\partial z / \partial x = 4$ and $\partial z / \partial y = 2$, and at $(0, 0)$, $\partial z / \partial x = 0$ and $\partial z / \partial y = 1$.

Problem 10.1

Use the basic rules of differentiation to find $\partial z / \partial x$ and $\partial z / \partial y$ for the functions $z = f(x, y)$, where $f(x, y)$ is given by

(a) $3x^2 + 2xy^3 - 3y^2$ (b) $(x - y)/(x + y)$ (c) $\cos(2x/y)$
(d) $(x^2 + y^2)e^{x^2 - y^2}$.

Problem 10.2

The entropy function for 1 mol of a monatomic ideal gas may be expressed in terms of V and T in the form

$$S = R \ln \left(\frac{e^{5/2} V}{L h^3} (2\pi m k T)^{3/2} \right),$$

where the symbols for the fundamental constants have their usual meaning, and L is the Avogadro constant.

(a) Given that $(\partial S / \partial T)_V = C_V / T$, where C_V is the heat capacity of the gas at constant volume, write down an expression for C_V.
(b) Using the equation of state for an ideal gas, $PV = RT$, express S in terms of T and P; thence, using the result that $(\partial S / \partial T)_P = C_P / T$, where C_P is the heat capacity at constant pressure, write down an expression for C_P.
(c) Show that $C_P / C_V = 5/3$.

Higher-order partial derivatives

As already noted, the partial derivatives of a given function, $f(x, y)$, are usually functions of both x and y, thereby making it possible to define higher-order partial derivatives by further partial differentiation with respect to either variable. This procedure results in second or higher partial derivatives occurring with mixed forms, such as

$$\frac{\partial^2 z}{\partial x \partial y} \left(\equiv \frac{\partial}{\partial x} \left(\frac{\partial z}{\partial y} \right) \right) \quad \text{or} \quad \frac{\partial^3 z}{\partial y^2 \partial x} \left(\equiv \frac{\partial}{\partial y} \left(\frac{\partial}{\partial y} \left(\frac{\partial z}{\partial x} \right) \right) \right), \quad \text{etc.}$$

▷

These include functions which describe physical properties.

In this procedure for obtaining higher-order derivatives, the order of partial differentiation with respect to each variable is of no significance for well-behaved functions ▷. In these, and other cases, the subscripts indicating which variables are held constant are suppressed for notational simplicity, when there is no chance of ambiguity. Thus, for example, in the third-order derivative above, the innermost and the two outer brackets would carry the subscripts y, x and x, respectively.

Worked example

10.2   For the function given in Example 10.1 find $\dfrac{\partial^3 z}{\partial x^2 \partial y}$ and $\dfrac{\partial^3 z}{\partial x \partial y \partial x}$.

Solution

$$\frac{\partial^3 z}{\partial x^2 \partial y} = \frac{\partial}{\partial x} \left(\frac{\partial}{\partial x} (x^2 + 1) \right) = \frac{\partial}{\partial x} (2x) = 2 \text{ and}$$

$$\frac{\partial^3 z}{\partial x \partial y \partial x} = \frac{\partial}{\partial x} \left(\frac{\partial}{\partial y} (2xy) \right) = \frac{\partial}{\partial x} (2x) = 2.$$

Problem 10.3

For each of the functions given in Problem 10.1, derive $\partial^2 z / \partial y \partial x$ and $\partial^2 z / \partial x \partial y$.

Problem 10.4

▷

The operator inside the outer round brackets is the spherical polar coordinate form of the operator

$$\frac{\partial^2}{\partial x^2} + \frac{\partial^2}{\partial y^2} + \frac{\partial^2}{\partial z^2}.$$

The Schrödinger equation for the hydrogen atom may be written in the form ▷:

$$-\frac{1}{2} \left(\frac{\partial^2 \psi}{\partial r^2} + \frac{2}{r} \cdot \frac{\partial \psi}{\partial r} + \frac{1}{r^2 \sin \theta} \cdot \frac{\partial}{\partial \theta} \left(\sin \theta \frac{\partial \psi}{\partial \theta} \right) \right.$$

$$\left. + \frac{1}{r^2 \sin^2 \theta} \cdot \frac{\partial^2 \psi}{\partial \phi^2} \right) - \frac{\psi}{r} = E\psi$$

where ψ is an atomic orbital wavefunction with energy E, and the position coordinates of the electron are given in spherical polar form as (r, θ, ϕ).

For the $2p_0$ atomic orbital $\psi = Nre^{-r/2}\cos\theta$ (N is a constant), show that ▷

$$\frac{\partial \psi}{\partial r} = N\left(1 - \frac{r}{2}\right)e^{-r/2}\cos\theta, \quad \frac{\partial \psi}{\partial \theta} = -Nre^{-r/2}\sin\theta,$$

$$\frac{\partial^2 \psi}{\partial r^2} = N\left(\frac{r}{4} - 1\right)e^{-r/2}\cos\theta, \quad \text{and} \quad \frac{\partial \psi}{\partial \phi} = 0.$$

Hence demonstrate that the left-hand side of the Schrödinger equation reduces to

$$-N\frac{r}{8}e^{-r/2}\cos\theta,$$

and give the value of E in eV ▷.

Differentiating under the integral sign – a useful procedure

Definite integrals often contain parameters, other than the integration variable in the integrand. When this is so, we can partially differentiate the integrand with respect to one of these parameters (assuming that these do not occur in either of the limits) to obtain a new definite integral whose value is obtained by partially differentiating the result obtained from the first integration. An example will make this procedure clear. Consider the integral

$$I = \int_0^\infty e^{-kr}dr = \left[\frac{e^{-kr}}{-k}\right]_0^\infty = \frac{1}{k},$$

the value of which depends on the parameter, k. Thus, partially differentiating with respect to k means that

$$\frac{\partial I}{\partial k} = \int_0^\infty (-r)e^{-kr}dr = -\frac{1}{k^2} \Rightarrow \int_0^\infty re^{-kr}dr = \frac{1}{k^2}.$$

Problem 10.5

Extend the procedure just described by evaluating the second, third and fourth partial derivatives of I with respect to k, and thereby deduce an expression for

$$\int_0^\infty r^n e^{-kr}dr.$$

In some instances, where the integrand does not contain an identifiable parameter like k, such a parameter can be introduced and, if the new integral is tractable, partial differentiation then follows as described above before resetting the parameter to unity.

> ▷
> Dwight, Equation 861.3
> with $a^2 = k$.

> **Problem 10.6**
>
> Given that ▷ $\int_0^\infty e^{-kr^2}\, dr = \frac{1}{2}\sqrt{\frac{\pi}{k}}$, show that $\int_0^\infty r^4 e^{-r^2}\, dr = \frac{3}{8}\sqrt{\pi}$.

Maxima, minima and saddle points

Functions of a single variable, for which the derivative is zero, yield turning points, which can be identified as maxima, minima or points of inflexion. For functions of two variables, however, the situation is similar, but more complicated. Points at which the two partial derivatives $\partial z/\partial x$ and $\partial z/\partial y$ are zero are termed *critical points*, and a test involving the three second partial derivatives is necessary to distinguish between a maximum, a minimum and a saddle point (in the latter situation, a maximum is achieved along one path on the surface, and a minimum along another path in a direction at 90° to the first path). The test involves first evaluating the three second-order partial derivatives $\partial^2 z/\partial x^2$, $\partial^2 z/\partial y^2$, $\partial^2 z/\partial x\partial y$, for values of x, y at the critical points; then the sign of

$$D = \frac{\partial^2 z}{\partial x^2} \cdot \frac{\partial^2 z}{\partial y^2} - \frac{\partial^2 z}{\partial x\partial y} \cdot \frac{\partial^2 z}{\partial x\partial y}$$

is investigated at each critical point. If $D > 0$, then either a minimum ($\partial^2 z/\partial x^2 > 0$), or a maximum ($\partial^2 z/\partial x^2 < 0$) is present; a saddle point occurs when $D < 0$ ▷. If $D = 0$ then further tests are required, just as in the analogous situation for functions of a single variable.

> ▷
> Salas and Hille, Section
> 16.7 – but note we have
> given a slightly different
> form of the test involving
> second-order partial
> derivatives.

> **Problem 10.7**
>
> (a) Verify that the location and nature of the critical points of the function $z = x^2 + 2y^2 - x^3$ are as given in the legend for Figure 10.1.
> (b) For the function $z = e^{-x}(x^2 - y^2)$, show that critical points are present at $(0,0)$ and $(2,0)$. Calculate the respective values of D, and determine the nature of each critical point.

The calculation of surfaces associated with the electronic states of dissociating or reacting molecular species plays an important role in elucidating the nature of transition states (saddle points) or the occurrence of different isomeric species (local minima). In the chemical situation, however, the energy functions of interest have more than two independent variables, and the topography of the resulting (hyper)surfaces is invariably complicated.

10.3   The differential, dz

In our consideration of the differential of a function of a single variable in Chapter 5, it was seen how the tangent approximation could be used to evaluate the change in the function value for a specific change in the independent variable. These ideas extend readily to functions $z = f(x, y, \ldots)$ of two or more variables, and have an important application within the thermodynamic context.

For a function of two variables, the change in z when x and y are incremented by Δx and Δy, respectively, is given by $\Delta z = f(x + \Delta x, y + \Delta y) - f(x, y)$. This change can be written in the form

$$\Delta z = \frac{\partial z}{\partial x} \Delta x + \frac{\partial z}{\partial y} \Delta y + \text{higher order terms in } \Delta x, \Delta y,$$

$\triangleright$
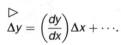
$$\Delta y = \left(\frac{dy}{dx}\right)\Delta x + \cdots.$$

and provides a generalization of the result for functions of a single variable $\triangleright$. The differential dz is defined as

$$dz = \frac{\partial z}{\partial x} dx + \frac{\partial z}{\partial y} dy \tag{10.1}$$

since, as for the single variable case, the differentials of the independent variables are given by the respective increments Δx and Δy. Just as for functions of a single variable, if the differentials of the independent variables are small in magnitude, dz provides a good approximation to Δz.

Problem 10.8

Write down the expression for

(a) dz for $z = xe^y + \ln(y/x)$
(b) dS for the entropy function defined in Problem 10.2 in terms of the variables T and P.

10.4   Application of differentials to error calculations

Formulae with a single measured property

Consider first the calculation of the error in a calculated property, y, when the formula used for its evaluation depends only upon a single measured variable, x. For the function $y = x^n$, the approximate change in y when x is subject to a change dx is

$$dy = \frac{dy}{dx} dx = nx^{n-1} dx.$$

$\triangleright$

The percentage error is found by multiplying the relative error by 100.

If dx is regarded as the error in x then, for small dx, dy gives a good approximation to the error in y. The *relative* error in y is defined as dy/y, and is seen to be given in this example as n times the relative error in x $\triangleright$. Thus, formulae involving integer powers of a measured quantity lead to a magnification of the error in the estimate for the derived property.

$\triangleright$

Alberty and Silbey, Section 8.11. See Problem 2.9(c).

> ### Problem 10.9
>
> The pH of a solution may be measured using a glass electrode with a typical error of ± 0.01. Calculate the estimated error in the hydrogen ion activity, a_{H^+}, when the measured pH is 1.0, given that $a_{H^+} = e^{-pH \ln 10}$.

Formulae with two or more measured properties

Consider a formula relating two variables, x, y, which may be subject to errors, to a property z. Just as for the single variable case, as noted above, if the measured errors are small, the differentials dx, dy may be used in estimating the error in z through the calculation of dz. The relative and percentage errors in z are given as above by dz/z and $dz/z \times 100$, respectively.

Before considering an example of error propagation, it is perhaps useful to note the extension of the expression for dz when there are n independent variables $x_1, x_2, \ldots x_n$:

$$dz = \frac{\partial z}{\partial x_1} dx_1 + \frac{\partial z}{\partial x_2} dx_2 + \cdots + \frac{\partial z}{\partial x_n} dx_n,$$

where $z = f(x_1, x_2, \ldots, x_n)$.

Worked example

10.3   The volume of a cylinder of height h, and cross-sectional radius r, is given by $V = \pi r^2 h$. If the measured values of r and h are 0.1 m and 1.0 m respectively, to within ± 0.02 m, what is the maximum percentage error in V?

Solution
$$dV = \frac{\partial V}{\partial r} dr + \frac{\partial V}{\partial h} dh \Rightarrow dV = 2\pi r h dr + \pi r^2 dh \Rightarrow \frac{dV}{V} = 2\frac{dr}{r} + \frac{dh}{h}.$$

Hence, for $dr = \pm 0.02$ m, $dh = \pm 0.02$ m, the maximum relative error in V,

$$\frac{2 \times 0.02}{0.1} + \frac{0.02}{1.0} = 0.42$$

▷
Notice that because V depends on the square of r, the error in r is magnified by a factor of two in the propagation process.

corresponds to a maximum percentage error of 42% ▷.

Problem 10.10

In the first-order rate process, in which A $\xrightarrow{k}$ B, the concentration of A (a, say) at time t is given by $a = \alpha e^{-kt}$, where $a = \alpha$ when $t = 0$. Assuming that α is known exactly, show that the magnitude of the maximum relative error in a is given by

$$kt\left(\frac{|dt|}{t} + \frac{|dk|}{k}\right).$$

Problem 10.11

The length of the hypotenuse, z, is related to the lengths x, y of the adjacent sides in a right-angled triangle according to the formula $z = \sqrt{x^2 + y^2}$. Determine the approximate change in length of the hypotenuse when the lengths of the adjacent sides change from 6.0 m and 8.0 m to 6.1 m and 8.1 m respectively.

10.5   The chain rule and the effects of changing variables

▷
Especially in thermodynamics.

In chemical situations ▷ it is often the case that we know the partial derivatives of a property in terms of one set of independent variables; the problem is then to use these derivatives to determine the corresponding partial derivatives with respect to the new variables.

In the case of two variables, if we are given $z = f(x, y)$ and both x and y are expressed in terms of two new variables, u and v, according to the given formulae $x = g(u, v)$, and $y = h(u, v)$, then we see that z may be expressed in terms of the variables u and v in the form $z = k(u, v)$. The partial derivatives $\partial z/\partial u$ and $\partial z/\partial v$ may be found without knowing the form of $k(u, v)$ as shown below.

From the given information, we have:

$$dz = \frac{\partial z}{\partial x} dx + \frac{\partial z}{\partial y} dy, \quad dx = \frac{\partial x}{\partial u} du + \frac{\partial x}{\partial v} dv, \quad dy = \frac{\partial y}{\partial u} du + \frac{\partial y}{\partial v} dv.$$

Substitution of the expressions for dx and dy into that for dz yields

$$dz = \left(\frac{\partial z}{\partial x} \cdot \frac{\partial x}{\partial u} + \frac{\partial z}{\partial y} \cdot \frac{\partial y}{\partial u} \right) du + \left(\frac{\partial z}{\partial x} \cdot \frac{\partial x}{\partial v} + \frac{\partial z}{\partial y} \cdot \frac{\partial y}{\partial v} \right) dv$$

$$= \frac{\partial z}{\partial u} du + \frac{\partial z}{\partial v} dv, \tag{10.2}$$

the last equality obtaining since $z = k(u, v)$. Hence, by comparing the terms multiplying du and dv in both equalities, it follows that

$$\frac{\partial z}{\partial u} = \frac{\partial z}{\partial x} \cdot \frac{\partial x}{\partial u} + \frac{\partial z}{\partial y} \cdot \frac{\partial y}{\partial u} \quad \text{and} \quad \frac{\partial z}{\partial v} = \frac{\partial z}{\partial x} \cdot \frac{\partial x}{\partial v} + \frac{\partial z}{\partial y} \cdot \frac{\partial y}{\partial v}.$$

Worked example

10.4   Find $\partial z/\partial u$, given that $z = 3x^2 \cos y + \ln x$ and $x = e^{u+v}$, $y = e^{u-v}$.

Solution   The four partial derivatives required in the construction of $\partial z/\partial u$ are

$$\frac{\partial z}{\partial x} = 6x \cos y + \frac{1}{x}, \quad \frac{\partial z}{\partial y} = -3x^2 \sin y, \quad \frac{\partial x}{\partial u} = e^{u+v} \text{ and } \frac{\partial y}{\partial u} = e^{u-v}.$$

Therefore:

$$\frac{\partial z}{\partial u} = \left(6x \cos y + \frac{1}{x} \right) e^{u+v} - 3x^2 e^{u-v} \sin y = 6e^{2(u+v)} \cos(e^{u-v})$$

$$+ 1 - 3e^{(3u+v)} \sin(e^{u-v}).$$

Problem 10.12

(a) Find $\dfrac{\partial z}{\partial v}$ for the function given in Example 10.4.

(b) If $z = x^2 + y^2$ and $x = s - 2t$, $y = 2s + t$, find $\dfrac{\partial z}{\partial s}$ and $\dfrac{\partial z}{\partial t}$.

10.6   Exact differentials

See Equation (5.8).

Suppose we have an expression of the form ▷

$$f(x, y)dx + g(x, y)dy \tag{10.3}$$

say, for example, $(2xy + 1)\,dx + (x^2 + 2)\,dy$, the question arises as to whether it corresponds to the differential of z as defined in Equation (10.1). If Equation (10.3) can be written as dz then, from Equation (10.1), we can identify $f(x, y)$ with $\partial z / \partial x$ and $g(x, y)$ with $\partial z / \partial y$.

Now we saw in Problem 10.3 above that

$$\frac{\partial^2 z}{\partial x \partial y} = \frac{\partial^2 z}{\partial y \partial x}, \text{ i.e. } \frac{\partial}{\partial x}\frac{\partial z}{\partial y} = \frac{\partial}{\partial y}\frac{\partial z}{\partial x},$$

illustrating the commutativity of $\partial / \partial x$ and $\partial / \partial y$ in all four cases: a result which is, in fact, true for all well-behaved (continuous) functions. Thus identifying $\partial z / \partial x$ with $f(x, y)$, and $\partial z / \partial y$ with $g(x, y)$, it follows immediately that the requirement for an expression of the form given in Equation (10.3) to be a differential is

$$\frac{\partial}{\partial x} g(x, y) = \frac{\partial}{\partial y} f(x, y). \tag{10.4}$$

In these circumstances, the function $z = F(x, y)$ may be defined, and its differential is given by

$$dz = \frac{\partial z}{\partial x} dx + \frac{\partial z}{\partial y} dy = f(x, y)dx + g(x, y)dy.$$

For historical reasons, the original expression in Equation (10.3) is then termed an *exact* differential. The previously cited expression $(2xy + 1)\,dx + (x^2 + 2)\,dy$ *is* an exact differential, since $f(x, y) = 2xy + 1$ and $g(x, y) = x^2 + 2$ and the requirement in Equation (10.4) is obeyed. If a differential is not exact then, as explained in the last section of Chapter 5, it is not strictly a differential at all, and it is as well to keep this in mind in order to avoid the confusion to which the usual terminology leads.

> **Problem 10.13**
>
> Identify which of the following is an exact differential:
>
> (a) $xy\,dx + y^2\,dy$ (b) $-(2x + y)dx + (y - x)dy$ (c) $e^y\,dx + (xe^y - 2y)dy$.

Finding the function, given its differential

For any differential expression of the form of Equation (10.3), the parent function $z = F(x, y)$ is readily found by performing two integrations. Consider, for example, the differential $dz = (2xy + 1)\,dx + (x^2 + 2)\,dy$, which was shown above to be an exact differential. Here

$$\frac{\partial z}{\partial x} = (2xy + 1), \quad \frac{\partial z}{\partial y} = (x^2 + 2),$$

▷
Notice the undetermined
function of y, $p(y)$, which
acts as a constant.

and integration of $\partial z/\partial x$, keeping y constant, yields ▷

$$z = F(x,y) = x^2y + x + p(y) + C\,. \tag{10.5}$$

Similarly, integration of $\partial z/\partial y$, keeping x constant, yields

$$z = F(x,y) = x^2y + 2y + q(x) + D\,. \tag{10.6}$$

Comparing Equations (10.5) and (10.6), we see that $p(y) = 2y$ and $q(x) = x$; thus,

$$z = F(x,y) = x^2y + x + 2y + A\,,$$

where A is a constant.

> ### Problem 10.14
>
> Find the function $F(x,y)$ for each of the exact differentials found in Problem 10.13.

▷
Traditionally called an
integrating factor.

As discussed in the concluding sections of Chapter 5, there are some physical properties, like work and heat, that cannot be represented by functions without further more detailed specification of how the work is done, or how the heat is transferred and so on. Thus, it is not possible to represent a change in these sorts of properties by a differential. It is, however, the custom to refer to small changes in these properties as non-exact differentials, as explained in Chapter 5. An example of such a non-exact differential is given in Problem 10.13(a) and in this case, and in some others of this simple form, it is possible to find a multiplicative factor ▷ that enables the given expression to be converted into a differential. Thus, in Problem 10.13(a) the given expression for the non-exact differential, dX, say, can be multiplied by a product of powers of x and y in the form $x^p y^q$. An integrating factor can then be sought by applying the test in Equation (10.4) to yield $(q + 1)x^2 = py^2$, after cancelling common factors. Since this equation is true for any choice for both x and y, it follows that $p = 0$ and $(q + 1) = 0$: that is, the integrating factor is $x^0 y^{-1} = 1/y$. Thus the expression dX/y defines an exact differential, $dX/y = dz = x\,dx + y\,dy$, which yields $z = F(x,y) = (x^2 + y^2)/2 + C$.

It is always possible to perform this kind of manipulation for a function of two variables. Note, however, that when the expression for the supposed differential of a function depends upon three or more variables, this method of determining an integrating factor is not guaranteed to work!

10.6   Thermodynamic applications

The thermodynamic quantities $P, U, S, H, G, \ldots$ etc, can be represented as functions of many variables, and thus all have differentials. In this context, these differentials are always called exact differentials, and the functions

▷
Atkins, Chapter 3.

themselves are always termed state functions. Work and heat, however, as already noted above, cannot be represented as functions, unless information is supplied about how the work is done, or how the heat is transferred; and this usually involves specifying the path along which the change is made ▷. Thus, the overall change in a given property, from state I to state II, is *independent* of the path from $I \longrightarrow II$ only if the property corresponds to a state function:

$$\int_I^{II} dX = X_{II} - X_I = \Delta X \,.$$

In other cases, involving work or heat, for example, the overall change in the property depends on the path selected, and is in no way connected with finding an integrating factor.

Problem 10.15

Using the equation of state for 1 mol of an ideal gas:
$$P = \frac{RT}{V} \equiv f(T, V) \,,$$

(a) write down the expression for dP and prove that it is an exact differential;

(b) obtain an expression for the work, W, done on the system undergoing isothermal compression from volume V_1 to V_2 under an externally applied pressure, P_{ex}, using the result

$$W = -\int_{V_1}^{V_2} P_{ex} \, dV \,,$$

when (i) the externally applied pressure is constant, and (ii) P_{ex} is always chosen so that the system is in equilibrium (that is, $P_{ex} = \frac{RT}{V}$).

It should now be apparent from the results obtained in the previous problem that work is not a state function, because $P_{ex} \, dV$ cannot be written as the differential of the work function, W.

Summary: In this second discussion of the differential calculus, we have extended ideas associated with functions of a single variable to functions described in terms of two or more variables – in preparation for more realistic applications within a chemical context. In the next chapter we conclude our discussion of the calculus of functions of many variables by considering the formulation of integration; the other important topics of power series and differential equations involving such functions are then easy enough to assimilate from more specialized texts using the formalism and concepts developed here.

11 Multiple integrals—integrating functions of several variables

This chapter

- extends the notion of integration to functions of two or more variables

- focuses on applications within the chemical context – especially in quantum chemistry, and in thermodynamics

▷
Because the coordinates in the problem always have initial and final values specified.

In Chapter 6 we approached the problem of integrating a function of a single variable by viewing integration as antidifferentiation. This provided us with several routes to determining indefinite integrals. When the integration variable is limited to the interval $[a, b]$, we are led to the definite integral which is, in fact, usually the realization of an integration process in the chemical situation ▷. The links between the two kinds of integration procedure were then established, and it was shown that the indefinite integral could be written as a definite integral with a variable upper limit – despite the fact that the actual integration itself is often easier to carry out by finding a suitable antiderivative.

The problem is that, as seen in Chapter 9, we frequently meet functions of two or more variables within the chemical context – simply because two external variables are usually needed to define the state of the system, in addition to any other variables, such as the concentrations of reacting species. For the moment, however, we focus on a simple mathematical example involving a function of two Cartesian coordinates.

11.1 Double integrals in terms of Cartesian coordinates

Suppose the given function is of the form $f(x, y) = x^2 + y^2$. If x is fixed and $f(x, y)$ is then integrated over the range of y coordinates from α_1 to α_2, the result is a function of the variable x, $g(x)$ say, where

$$\int_{\alpha_1}^{\alpha_2} f(x, y) dy = g(x).$$

For a given x, the value of $g(x)$ also depends upon the values of the end-points α_1 and α_2, which are constants. Further integration over the values of x contained within the interval $[\beta_1, \beta_2]$ then yields the result

$$\int_{\beta_1}^{\beta_2} g(x)dx = \int_{\beta_1}^{\beta_2} \left(\int_{\alpha_1}^{\alpha_2} f(x,y)dy \right) dx \,,$$

the value of which depends only upon the two sets of end-points. The brackets are used to identify the sequence of integrations which, in all normal situations, is of no significance. The result is therefore independent of the order of integration, and the brackets may be dropped:

$$\int_{\beta_1}^{\beta_2} \int_{\alpha_1}^{\alpha_2} f(x,y)dx \, dy \,.$$

If, however, it is necessary to indicate the order of integration unambiguously, then it is common practice to write the double integral in either of the alternative forms

$$\int_{\alpha_1}^{\alpha_2} dy \int_{\beta_1}^{\beta_2} dx \, f(x,y) \quad \text{or} \quad \int_{\beta_1}^{\beta_2} dx \int_{\alpha_1}^{\alpha_2} dy f(x,y) \,, \tag{11.1}$$

which do not involve the explicit use of brackets.

Each equivalent form of the double integral in Equation (11.1) involves integrating $f(x,y)$ over a rectangular region with vertices given by (β_1, α_1), (β_1, α_2), (β_2, α_1) and (β_2, α_2), as shown in Figure 11.1. In other instances (as considered later), the region over which the integration takes place may have a different shape, or it might be the whole xy plane, when $\alpha_1, \beta_1 \to -\infty$ and $\alpha_2, \beta_2 \to \infty$.

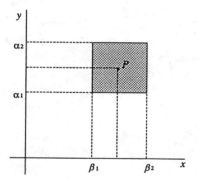

Figure 11.1
The rectangular region for a double integral

As far as the integrations in themselves are concerned, sometimes it is easier to integrate over the y variable first, and sometimes second: but, as already indicated, for a double integral associated with a physical problem, the result has to be the same irrespective of the sequence of integrations. The next example illustrates this situation.

Worked example

11.1   Evaluate

$$\int_0^1 \int_0^1 (x^2 + y^2)dx\,dy\,.$$

Solution   Integrating over x first yields

$$\int_0^1 \left(\int_0^1 (x^2 + y^2)dx \right) dy = \int_0^1 \left[\frac{x^3}{3} + y^2 x \right]_0^1 dy = \int_0^1 \left(\frac{1}{3} + y^2 \right) dy$$

$$= \left[\frac{y}{3} + \frac{y^3}{3} \right]_0^1 = \frac{2}{3}\,,$$

whilst integrating over y first yields

$$\int_0^1 \left(\int_0^1 (x^2 + y^2)dy \right) dx = \int_0^1 \left[x^2 y + \frac{y^3}{3} \right]_0^1 dx = \int_0^1 \left(x^2 + \frac{1}{3} \right) dx$$

$$= \left[\frac{x^3}{3} + \frac{x}{3} \right]_0^1 = \frac{2}{3}\,.$$

Problem 11.1

For the function $f(x, y) = xe^{xy}$,
(a) integrate over y first, and show that

$$I = \int_0^1 dx \int_0^1 xe^{xy}\,dy = \int_0^1 dx\,x \int_0^1 e^{xy}\,dy = e - 2\,;$$

(b) integrate over x first and then, using integration by parts, show that

$$I = \int_0^1 dy \int_0^1 xe^{xy}\,dx = \int_0^1 \left(\frac{e^y}{y} - \frac{e^y}{y^2} + \frac{1}{y^2} \right) dy$$

$$= \lim_{\epsilon \to 0} \left\{ \int_\epsilon^1 \frac{e^y}{y}\,dy - \int_\epsilon^1 \frac{e^y}{y^2}\,dy + \int_\epsilon^1 \frac{1}{y^2}\,dy \right\},$$

where, in the last step, the limit is necessary because each of the three integrands becomes infinite at the lower limit ▷;
(c) rewrite the second integral using integration by parts, evaluate the third integral and thus derive the result

$$I = \lim_{\epsilon \to 0} \left\{ \int_\epsilon^1 \frac{e^y}{y}\,dy + \left[\frac{e^y}{y} \right]_\epsilon^1 - \int_\epsilon^1 \frac{e^y}{y}\,dy - \left[\frac{1}{y} \right]_\epsilon^1 \right\}$$

$$= e - 1 - \lim_{\epsilon \to 0} \left(\frac{e^\epsilon}{\epsilon} - \frac{1}{\epsilon} \right);$$

▷
See Chapter 6.

(d) use the Maclaurin series for e^{ϵ} in the form $e^{\epsilon} = 1 + \epsilon + \cdots$, and demonstrate that the value of $e - 2$ for I is regained for the value of the double integral $\triangleright$.

▷
But by much more effort than in part (a).

The choice of a rectangular region for carrying out the two integrations considered so far reflects the use of a Cartesian coordinate system, insofar as for *any* choice of x the permitted range of y values is $[\alpha_1, \alpha_2]$. Integration over polygonal or annular $\triangleright$ regions, or sectors, results in the range of permitted y values depending upon the choice of x, and this necessitates some care in performing the integrations.

▷
A ring-shaped region, delineated by two circular arcs as boundaries as shown in Figure 11.5.

Integration over a triangular region

Consider, for example, the double integral of $f(x,y) = xy + 1$ over the triangular region defined by the lines $y = 1, y = x + 1$ and $x = 1$ shown in Figure 11.2. If we integrate over y first then, as x takes values from 0 to 1, the values of y are constrained to lie between 1 $(x = 0)$ and the value $x + 1$ $\triangleright$. Hence, one way of writing the integral is

▷
The equation of the straight line defining the upper limit for y is
$y = x + 1$.

$$\int_0^1 dx \int_1^{x+1} (xy + 1)\, dy. \tag{11.2}$$

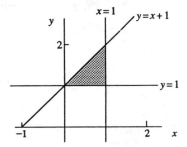

▷

Figure 11.2
The triangular region for the integral in Equation (11.2)

On the other hand, if we integrate over x first then, for a fixed value of y in the interval $[1,2]$, the permitted values of x range from $x = y - 1$ to 1, yielding the integral in the form

$$\int_1^2 dy \int_{y-1}^1 (xy + 1)\, dx. \tag{11.3}$$

Integration over a sector

The double integral of $f(x, y)$ over a sector is approached first by using Cartesian coordinates and then working out the upper and lower limits on x and y following the procedure described for the triangular region.

> **Problem 11.2**
>
> Evaluate
>
> (a) the integrals given in Equations (11.2), (11.3), and show that the same value is obtained in both cases.
> (b) the integral of $f(x, y) = xy + 1$ over the region bounded by the circle $x^2 + y^2 = 1$, and the positive x and y axes ▷, and verify that the same result is obtained irrespective of whether y is integrated first or second.

▷

The values of x and y are such that $0 \le x \le 1$ and $y > 0$, with the upper limit on y determined by the equation of the circle.

The last part of Problem 11.2 showed that, although the integral in terms of Cartesian coordinates could be performed over a sector (in general an annular region), the analysis is more involved than when dealing with rectangular regions. The reason for this is that the Cartesian coordinate system is inappropriate: plane polar coordinates r, θ form a better choice when dealing with annular regions, as they are more naturally predisposed towards integrating over regions bounded by circular arcs. The problem is, how do we rewrite the integral in terms of plane polar coordinates? The clue to resolving this issue is to notice that, when Cartesian coordinates are used, the product of the differentials, $dy\,dx$, appears in the integral and this corresponds to the element of area generated when, for a point $P(x, y)$ (see Figure 11.3), x is incremented by dx, and y by dy. For plane polar coordinates, the element of area generated when the coordinates r and θ are incremented by dr and $d\theta$, respectively, is given by $r\,dr\,d\theta$ ▷, as can be seen in Figure 11.4.

▷

The Jacobian for the transformation from Cartesian to plane polar coordinates, as seen later in Chapter 13.

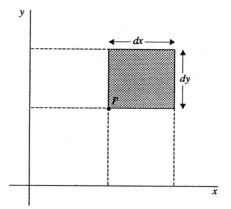

Figure 11.3
The element of area in Cartesian coordinates

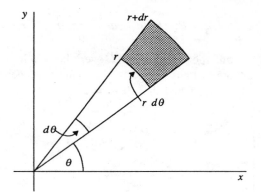

Figure 11.4
The element of area in plane
polar coordinates

The function $f(x,y)$ forming the integrand of the double integral now becomes a function of r, θ in the form $F(r, \theta) = f(r \cos \theta, r \sin \theta)$, after making the usual substitutions for x, y. It is clear, moreover, that $f(x, y)$ and $F(r, \theta)$ cannot describe the same function as the two formulae, with their differing domains, are clearly not the same. Thus, the double integral of $f(x, y)$ over some region Γ, bounded by one or two circular arcs, may be written as $\int_{\Gamma} \int f(x, y) \, dx \, dy$ in Cartesian coordinates or as $\int_{\Gamma} \int F(r, \theta) \, r \, dr \, d\theta$ where, of course, the actual values for the limits for the integration variables depend upon the choice of coordinate system.

Worked example

11.2   Use plane polar coordinates to evaluate the integral in Problem 11.2 over the region bounded by the circular arc as specified in the problem.

Solution   The function $F(r, \theta)$ is given by $r^2 \cos \theta \sin \theta + 1$ and the integral required is

$$\int_0^1 r \, dr \int_0^{\pi/4} d\theta (r^2 \cos \theta \sin \theta + 1) = \int_0^1 r^3 \, dr \int_0^{\pi/4} \cos \theta \sin \theta \, d\theta$$

$$+ \int_0^1 r \, dr \int_0^{\pi/4} d\theta$$

$$= \left[\frac{r^4}{4} \right]_0^1 \left[\frac{1}{2} \sin^2 \theta \right]_0^{\pi/4} + \left[\frac{r^2}{2} \right]_0^1 [\theta]_0^{\pi/4} = \frac{1}{8} + \frac{\pi}{8} .$$

Integration over an annular region

The integration of $f(x, y)$ over a general annular region shown in Figure 11.5 is readily carried out by forming $F(r, \theta)$ and then evaluating

$$\int_{r_1}^{r_2} \left(\int_{\theta_1}^{\theta_2} F(r, \theta) d\theta \right) r \, dr = \int_{r_1}^{r_2} \int_{\theta_1}^{\theta_2} F(r, \theta) r \, dr \, d\theta \, .$$

If, instead, the integration is over the whole xy plane, then it may be appropriate to transform $f(x, y)$ to plane polar coordinates and use the limits $r_1 = 0, r_2 \to \infty$, and $\theta_1 = 0, \theta_2 = 2\pi$.

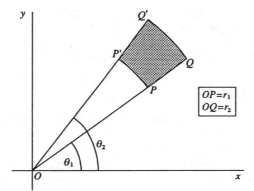

Figure 11.5
An annular element of area
in the xy plane

> **Problem 11.3**
>
> Integrate the function given in Example 11.2 over the annular region defined by $\theta_1 = \pi/4$, $\theta_2 = \pi/3$, $r_1 = 1$, $r_2 = 2$.

11.3   **Worked example**

Solution   Evaluate $\displaystyle\int_{-\infty}^{\infty} \int_{-\infty}^{\infty} e^{-\sqrt{x^2+y^2}} \cdot \frac{x^2}{x^2+y^2} \, dx \, dy$.

This integration over the entire xy plane is best evaluated by first transforming to plane polar coordinates. The function $F(r, \theta)$ then becomes $e^{-r} \cos^2 \theta$ and, in this case, the double integral splits into a product of integrals – one over each variable in the form ▷

▷
Remember that the integral sign with the infinite upper limit is a shorthand for

$$\lim_{L \to \infty} \int_0^L$$

$$\int_0^\infty r \, dr \int_0^{2\pi} d\theta \, e^{-r} \cos^2 \theta = \left[\int_0^\infty e^{-r} r \, dr \right] \left[\int_0^{2\pi} \cos^2 \theta \, d\theta \right]$$
$$= 1 \times \pi = \pi \, ,$$

where the integration over r is carried out using integration by parts (see also Problem 10.5), and the double angle formula is used to facilitate the integration over θ. Transforming to plane polar coordinates simplifies matters considerably in this problem; however, integrations over finite rectangular regions are best carried out using Cartesian coordinates.

Problem 11.4

Evaluate the double integral of $f(x,y)$ over

(a) the region bounded by the lines $x = 1, x = 3, y = 2, y = 4$, where $f(x,y) = xy$,
(b) the region bounded by the semicircle $x^2 + y^2 = 1$, $y \geq 0$, where $f(x,y) = xy$,
(c) the region bounded by the semicircle $x^2 + y^2 = 1$, $y \leq 0$, where $f(x,y) = (x^2 + y^2)^{\frac{1}{2}} e^{-(x^2+y^2)^{\frac{1}{2}}}$, using plane polar coordinates, and the value for the radial integral as found in Problem 10.5.

11.3   A special integral

In our consideration of the integration of functions of a single variable, we studiously avoided discussing in detail integrals involving an integrand of the form $f_n(x) = x^{2n} e^{-ax^2}$, where n is a non-zero integer (see Problem 10.6). Definite integrals of functions of this form occur in statistics (Chapter 12), and also in calculations of molecular orbital wavefunctions for molecules using so-called Gaussian atomic functions – just to give two examples of their occurrence within a chemical context.

Consider the double integral of $f_0(x) f_0(y)$ over the first quadrant in the xy plane:

$$M = \int_0^\infty dx \int_0^\infty dy\, e^{-a(x^2+y^2)} = \left(\int_0^\infty dx\, e^{-ax^2} \right) \left(\int_0^\infty dy\, e^{-ay^2} \right) = I^2,$$

▷
The integration over the y variable also yields I, as the value of a definite integral cannot depend upon the name of the integration variable.

where I is the value of the integral obtained by integrating over the x variable ▷. The double integral M is evaluated very simply by first transforming to plane polar coordinates:

$$M = \int_0^\infty r\, dr \int_0^{\pi/2} d\theta\, e^{-ar^2} = \frac{\pi}{2} \int_0^\infty dr\, r\, e^{-ar^2},$$

and then carrying out the integration over r by using the substitution $u = ar^2$ to obtain $M = \pi/4a$. Thus, given that $M = I^2$, we obtain $I = \left(\frac{\pi}{4a} \right)^{\frac{1}{2}} = \frac{1}{2} \sqrt{\frac{\pi}{a}}$.

▷
pp. 488–9

The next problem is taken from Berry, Rice and Ross, ▷ and a more complete explanation of the context is given in their text. Although we are concerned here with motion in three dimensions, one of the variables is removed from the problem through the way in which it is posed.

Problem 11.5

Consider a monatomic gas confined to a vessel with volume V. The pressure exerted on the wall, S, results from the net change of momentum perpendicular to the wall per unit time per unit area (see Figure 11.6).

The contribution to the pressure, P, from atoms with speeds between v and $v + dv$ is:

$$dP = 2mv \cos \theta \cdot F(\theta, v) \, d\theta \, dv$$

where $2\,mv \cos \theta$ is the change in momentum in the z-direction (see Fig. 11.6), and $F(\theta, v) \, d\theta \, dv$ is the number of atoms per unit time per unit area that strike S with speed between v and $v + dv$, and θ lying between θ and $\theta + d\theta$.

Then

$$P = \int dP = 2 \int_0^\infty \int_0^{\pi/2} mv \cos \theta \, F(\theta, v) \, d\theta \, dv$$

$$= \frac{m}{3} \int_0^\infty v^2 f(v) dv \,,$$

using the form of $F(\theta, v) = \frac{1}{2} v \cos \theta f(v) \sin \theta$ given by Berry, Rice and Ross. Given that $\triangleright$

$\triangleright$
Equation (12.11) of
Berry *et al.*

$$\int_0^\infty v^2 f(v) dv = \frac{2U}{mV} \,,$$

where U is the internal energy of the system,

(a) show that for 1 mol of a monatomic gas,

$$P = \frac{2}{3} \frac{U}{V} \,, \qquad U = \frac{3}{2} RT \,, \qquad \left(\frac{\partial U}{\partial T} \right)_V = \frac{3}{2} R \,, \qquad \left(\frac{\partial U}{\partial V} \right)_T = 0;$$

(b) give the name of the property represented by $(\partial U / \partial T)_V$.

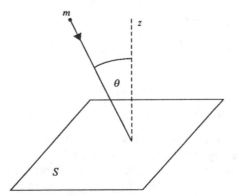

Figure 11.6
The coordinates defining the
angle of approach of an
atom to the wall S

11.4   Integrals involving functions with more than two variables

▷

Apart from other
coordinates such as time
or spin.

▷

$(3N - 5) + 3n$ for a
linear molecule.

▷

Ψ is assumed to be real
here.

Further applications of the techniques described here are to be found in kinetic theory; theories of electronic structure and of liquids; spectroscopy; etc. However, an extension of the methodology and notation is required because, as noted above, any particle moving in three dimensions needs three coordinates to specify its position in space ▷. This means that the (wave)function describing the system of N nuclei and n electrons depends upon $3N + 3n$ spatial coordinates; however, if only vibrations of the nuclei are of interest, then six coordinates can be discounted for a non-linear molecule, as the three rotations and three translations of the nuclei do not change the relative positions of the nuclei. With this assumption, the function, Ψ, describing the electronic motion in the presence of the vibrating nuclei depends upon $(3N - 6) + 3n$ spatial coordinates ▷. The integrations required, in working out the value of any (non-linear) molecular property, must therefore be carried out over all $(3N - 6)$ vibrational variables, s_i, and n sets of electronic coordinates $\mathbf{x}_i = (x_i, y_i, z_i)$. Since it is very tedious to write down all the $(3N - 6) + 3n$ integral signs, each with their respective limits, various symbolic notations are used. Thus, for example, the integral of the square ▷ of the wavefunction may be given in the form

$$\int \Psi \Psi \, d\mathbf{s} \, d\mathbf{x} ,$$

to indicate integration over all $3N - 6$ variables $s_1, s_2, \ldots, s_{3N-6}$ and all $3n$ electronic variables $(x_1, y_1, z_1, \ldots, x_n, y_n, z_n)$.

Problem 11.6

Consider the HCl molecule in its ground electronic and vibrational states, with an internuclear separation of R. Under the influence of electromagnetic radiation in the infrared region, energy is absorbed by the bond-stretching vibrational motion (changes in rotational energy are ignored for simplicity). Thus, for the ground and vibrationally excited states, $\Psi_0 = \psi_0(\mathbf{x}, R)\chi_0(s_1)$ and $\Psi_1 = \psi_0(\mathbf{x}, R)\chi_1(s_1)$, respectively, the intensity of the infrared transition between Ψ_0 and Ψ_1 is proportional to the square of the dipole moment integral ▷

▷

Atkins, Section 17.1

$$-e \int \Psi_0 \{z_1 + z_2 + \cdots + z_n\} \Psi_1 \, d\mathbf{x} \, ds_1$$

$$= -e \int \psi_0(\mathbf{x}, R)\chi_0(s_1)\{z_1 + z_2 + \cdots + z_n\}\psi_0(\mathbf{x}, R)\chi_1(s_1) \, d\mathbf{x} \, ds_1 ,$$

▷

The z-axis coincides with
the internuclear axis.

where z_i is the z-coordinate ▷ of the electron labelled i, and s_1 is the bond extension $R - R_e$. Integration over all $3n$ electron coordinates yields an electronic dipole function, $\mu(R)$, which depends upon R. The remaining integration yields the vibrationally averaged dipole moment in the form

$$\int \chi_0(s_1)\mu(R)\chi_1(s_1)\,ds_1 \, .$$

(a) Use the Taylor series expansion of $\mu(R)$ about R_e ▷, and the result that the two lowest vibrational wavefunctions for the ground electronic state do not overlap ($\int \chi_0(s_1)\chi_1(s_1)ds_1 = 0$), to show that an infrared transition is observed only if the derivative of the dipole moment function does not vanish.

(b) Repeat the analysis for electronic excitation, where $\Psi_0 = \psi_0\chi_0$ ▷, and $\Psi_1 = \psi_1\chi_0'$, and $\mu(R)$ is now the transition dipole $-e\int \psi_0(\mathbf{x}, R)\chi_0(s_1)\{z_1 + z_2 + \cdots + z_n\}\psi_1(\mathbf{x}, R)\chi_0' \, d\mathbf{x}ds_1$, to show that, when $\mu(R) \neq 0$, the intensity of the transition depends upon the square of the overlap between χ_0 and χ_0'.

▷
See Chapter 7.

▷
The prime on the lowest energy vibrational wavefunction in the excited electronic state indicates that it is not the same function as in the ground state.

▷
The problem is that the analytical form of the function describing the interaction is not known it is guessed!

Similar kinds of multiple integral to those arising in the last problem occur in the modelling of interacting molecular species, for example, in determining the equation of state for a system consisting of gaseous molecules, where the internal energy consists not only of the usual translational, rotational and vibrational contributions but also the effects of molecule–molecule interactions. Such interactions are usually assumed to be additive in a pairwise manner, and the evaluation of this contribution to the system energy involves the evaluation of multiple integrals ▷.

Summary: This concludes our exploration of how functions of many variables can be integrated – largely within a chemical context. As we have seen, it is important in this development to become familiar with the symbolic way of writing multiple integrals when there are many variables over which the integration is to be performed. It is the case that all authors of physical chemistry textbooks use the convention (often without comment) that the integral sign is interpreted to mean that the integration is to be carried out over all variables in the problem, and sometimes over the complete variable ranges; also, the differential associated with the integral sign is also written symbolically. This shorthand approach avoids the notation obscuring the basic message contained within the mathematics.

The next chapter provides a survey of the language and methodology of statistics. There is a strong link to the calculus of functions of several variables, and the chapter provides the basis for handling experimental data – in particular their fitting to an assumed formula.

12 Statistics

Objectives

This chapter

- shows how statistical ideas can be used to determine the reliability of a straight line fitted to experimental data

- develops these ideas to introduce the notion of statistical inference

- considers some particular cases of statistical inference useful in chemistry

12.1 Statistics in a chemical context

One of the first places in which chemists come across the need for statistical analysis is when asked to determine the order of a given reaction from measurements of reactant concentration made at various times. It is usual to postulate, on mechanistic grounds, a possible form for the rate of the reaction as a function of the concentration of the reactants. Thus if it is supposed that a reaction is first order in the concentration a of a reactant A, then the theoretical relationship between the rate of decrease of A and the time is:

$$-\frac{da}{dt} = ka$$

where k is a constant, usually called the first-order rate constant, which has units of s^{-1}. Naturally there will be other differential equations for other schemes but, as it is very difficult to do anything useful with the differential equation itself, it is generally integrated and the integrated form compared with experiment. A number of examples have already been encountered in Chapter 9, and the present case comprises Problem 9.2 where it is shown that the integrated form of the first-order equation may be written as:

$$\ln\frac{a}{\alpha} = -kt$$

where α is the concentration of the reactant A at $t = 0$.

In practice the concentration of A, a_i, is measured at discrete times t_i and we would try plotting $\ln \dfrac{a_i}{\alpha}$ against t_i. If we obtained a decent straight line, then we would believe that the assumption of first-order kinetics was correct and identify the slope of the line with the negative of the rate, k.

The points will not lie exactly on a straight line and so there is a measure of uncertainty about whether the first-order assumption is a soundly based one. Suppose, however, that we were satisfied that the straight line relationship was the proper one. Then, since a number of seemingly satisfactory straight lines can be drawn through the set of points, there will be a number of plausible values calculable for k. In what value or values might we have the most confidence? These problems are best tackled by statistical methods. Indeed nowadays most of us are probably first made aware of the statistical aspect of treating this sort of thing because programs to calculate 'the best straight line fit' are now ubiquitous and perhaps more often used than plotting graphs by hand. The output of a run contains statistical information and it is not always clear how to interpret it. In this chapter we shall try to show how such interpretation should be made. We will do so in the context of a general straight line fit; an alternative approach to this problem is presented in Chapter 17.

12.2   The theory of linear regression

Let it be assumed that we have reason to believe that values of an independent variable, x, are linearly related to those of a (measured) dependent variable, y. Both x and y generally lie in the full domain $(-\infty, +\infty)$ but in some cases, only in a subdomain. The former is assumed unless otherwise specified.

In attempting to fit a straight line of the form $y = bx + a$ to our experimental data, which is exemplified by n measurements (x_i, y_i), $i = 1, 2, 3, \ldots, n$, we have the problem of determining a and b. If a and b were known, then for each value x_i we could compute a value of y, call it $\hat{y}_i$, and if the fit were exact then $(y_i - \hat{y}_i)$ would vanish. Of course, for a given set of data, a and b are generally unknown but if there is such a linear relationship then it would be plausible to *determine* a and b from the condition that the distance between y_i and $\hat{y}_i$ be minimized for all the points together. This is done by the method of *least squares* (sometimes called least *mean* squares). This method is described in slightly different terms in Chapter 17 but, for present purposes, we shall present it as the method which minimizes the quantity:

$$U(a,b) \equiv U = \sum_{i=1}^{n} (y_i - \hat{y}_i)^2 \tag{12.1}$$

▷
See Chapter 4.

with respect to variations in a and b. Substituting for $\hat{y}_i$, U becomes
$U = \sum_{i=1}^{n} (y_i - bx_i - a)^2$, and the conditions for the turning points ▷ are

$$\frac{\partial U}{\partial a} = \sum_{i=1}^{n} 2(y_i - bx_i - a)(-1) = 0, \qquad \frac{\partial U}{\partial b} = \sum_{i=1}^{n} 2(y_i - bx_i - a)(-x_i) = 0$$

$$\Rightarrow \sum_{i=1}^{n} y_i = \sum_{i=1}^{n} (bx_i + a), \qquad \sum_{i=1}^{n} x_i y_i = \sum_{i=1}^{n} (bx_i^2 + ax_i). \tag{12.2}$$

By defining the *arithmetic mean* (commonly called the *average* value) of the observations as

$$\bar{x} = \frac{1}{n} \sum_{i=1}^{n} x_i, \qquad \bar{y} = \frac{1}{n} \sum_{i=1}^{n} y_i$$

▷
The rewritten equations
are often called the
normal equations.

then we can rewrite Equations (12.2) ▷ as

$$n\bar{y} = nb\bar{x} + na \Rightarrow \bar{y} = b\bar{x} + a, \qquad \sum_{i=1}^{n} x_i y_i = na\bar{x} + b \sum_{i=1}^{n} x_i^2.$$

Eliminating a between these two equations gives for b:

$$b = \frac{\sum_{i=1}^{n} (x_i - \bar{x})(y_i - \bar{y})}{\sum_{i=1}^{n} (x_i - \bar{x})^2}. \tag{12.3}$$

Clearly if the value of the denominator in the above equation is vanishingly small then the solution becomes problematic because of numerical instability. However, assuming that all is well, then the appropriate value for a is

$$a = \bar{y} - b\bar{x} \tag{12.4}$$

where b is given by Equation (12.3).

The objective of the next problem is to show that, using the techniques described in Chapters 4, 5, 7, and 10, the above choices for a and b yield a turning point of U that corresponds to a minimum.

▷
For the definition of D,
see the section on
maxima and minima in
Chapter 10.

Problem 12.1

Show that

(a) $\dfrac{\partial^2 U}{\partial a^2} = 2n$, $\dfrac{\partial^2 U}{\partial b^2} = 2 \sum_{i=1}^{n} x_i^2$, $\dfrac{\partial^2 U}{\partial a\, \partial b} = 2 \sum_{i=1}^{n} x_i$

(b) ▷ $D = 4n \sum_{i=1}^{n} x_i^2 - 4 \left(\sum_{i=1}^{n} x_i \right)^2 = 4n^2 (\overline{x^2} - \bar{x}^2)$

(c) $\sum_{i=1}^{n} (x_i - \bar{x})^2 = n(\overline{x^2} - \bar{x}^2)$

$$\text{(d) } D = 4n \left(\sum_{i=1}^{n} (x_i - \bar{x})^2 \right) > 0.$$

Substituting for a in the straight line equation using Equation (12.4), it follows that the assumed linear relationship takes the form

$$(y - \bar{y}) = b(x - \bar{x}),$$

showing that if the relationship is properly given, then the regression line must pass through the means of both sets of points.

The calculations necessary to determine a and b by hand are rather tiresome but really are no longer necessary because not only computers and personal computers, but many pocket calculators, are programmed to do them. The major problem is in ensuring that the data points are entered accurately as input. Although we shall later provide some numerical problems in this area we assume that machines will be used to calculate the answers.

Worked example

▷   **12.1**   For the gas phase thermal decomposition of azomethane, $CH_3N_2CH_3$, at
Atkins, Example 25.4.   600 K ▷

$$CH_3N_2CH_3 \longrightarrow (CH_3)_2 + N_2$$

the following table gives the data for the partial pressure, p, of azomethane versus time, t.

t/s	$10^2 p/\text{mmHg}$
0	8.2
1000	5.72
2000	3.99
3000	2.78
4000	1.94

(a) Show that the reaction is first-order in azomethane.
(b) Determine the rate constant with respect to the disappearance of azomethane

Solution   (a) From our previous discussion, if the reaction is first-order, then a plot of the natural logarithm of the ratio of pressure to the initial pressure as dependent variable, against time as the independent variable, should give a straight line. The slope of this line is then the negative of the rate constant.

With the data above the calculated straight line appears to fit the observed points very well indeed, to give a slope (units s^{-1}) of -3.604×10^{-4} and an

▷
A widely available
package.

intercept (dimensionless) of 1.545×10^{-4}. This last value is effectively zero, as it should be.

(b) $k = -3.604 \times 10^{-4}$ s^{-1}

If these results had been obtained using the SPSS statistical package ▷ an entry in the output would read 'Multiple $R1.0000$'. There would be similar lines of output in other packages. We shall explain in the next section how this should be interpreted.

Problem 12.2

In Example 12.1

(a) how would changing the units in which pressure is given to SI units affect the results of the regression?

(b) using the definition of b given in Equation (12.3), explain how b would change if the independent variable were taken simply to be the logarithm of the pressure in a particular system of units

---

Now we turn to a consideration of how we can test whether what we have done is meaningful and, if meaningful, what kind of confidence we might ascribe to the calculated values of a and b.

12.3 Validating linear regression

As mentioned above, when looking at the output of a regression program it is usual to see quoted a statistic r, sometimes called the *correlation coefficient* and sometimes (more rarely) written R. This coefficient is defined as

$$r = \frac{\sum_{i=1}^{n}(x_i - \bar{x})(y_i - \bar{y})}{s_x s_y}$$

where s_x and s_y are obtained from the quantities

$$s_x^2 = \sum_{i=1}^{n}(x_i - \bar{x})^2 \qquad s_y^2 = \sum_{i=1}^{n}(y_i - \bar{y})^2 . \tag{12.5}$$

Using Equation (12.3) above, the expression for r can be rewritten as

$$r = b\frac{s_x}{s_y} . \tag{12.6}$$

If x and y are independent of each other, then b would be zero and r would vanish. On the other hand, if the points were to lie exactly on the straight line then $y_i = a + bx_i$ and $\bar{y} = a + b\bar{x}$. Thus,

$$\sum_{i=1}^{n}(y_i - \bar{y})^2 = b^2 \sum_{i=1}^{n}(x_i - \bar{x})^2 \Rightarrow s_y^2 = b^2 s_x^2 \,.$$

In this situation,

$$r^2 = b^2 \left(\frac{s_x}{s_y}\right)^2 = 1$$

and $r = \pm 1$ for an exact fit.

It is seen from Equation (12.6) that the sign of r is determined by the sign of b and so $r = -1$, for example, would signify an exact fit in the case of a line with negative slope. It is also the case that in the output from some programs (SPSS is one) $|r|$ is quoted, so that only the interval $[0, 1]$ is significant.

It thus seems reasonable to assert that if r is close to zero then there is, in fact, no linear relationship between the variables x and y. Notice that this does *not* mean that there is no relationship at all. Nothing can be said about that generally from this statistic. It simply means that even if there is a relationship, it is not a linear one. Similarly if $|r|$ is close to unity then there probably is a linear relationship between x and y. Notice too that in the absence of other evidence it is not safe to assert any causal relationship between the attributes represented by the variables. Statisticians have great fun by correlating alcoholic beverage consumption in the UK with the salaries of UK university teachers, or the number of storks in Denmark with the Danish birth rate. Many more cases like this, in which the values of $|r|$ are really quite large, could be quoted. They are perfectly useless in supporting any rational argument for a causal relation.

It would be logical at this point to investigate how we should decide when to accept a particular value of r as indicative of a linear (though not necessarily causal) relationship. Unfortunately to do this for the statistic r is rather a tricky problem and so we shall not consider it in detail. Later, however, we shall be able to show how, in principle, it could be done. To do both that, and other useful things, we shall need to make a digression to develop the idea of a *population* and also to introduce some rather typical kinds of statistical arguments. Since the experiment, implicit in the linear relationship, may be repeated again and again to obtain information at more points (x, y), it is possible to imagine that there is a *population* of such points. The set of points that we have actually taken is simply a sample drawn from that population, and this idea is now investigated further.

The distribution of the measured values

Suppose that for fixed x_i we made repeated measurements of y. Let us denote these as $y_{i,j}$ $(j = 1, 2, 3, \ldots, m)$, to emphasize that it is measurements for a particular value of x that are of interest. We could calculate a mean value $\bar{y}_i$ for this sample of measurements. A *median value* could also be calculated for the sample, such that half the measured points have values less than the median and the other half have values that are greater than the median. It could be that some of the $y_{i,j}$ have the same value or, if not exactly the same value, then values very close to each other. In such a case it is possible to choose a small range of y, and to count up the number of values that lie in that interval. If this is carried out for intervals to cover the whole range of y then (assuming equal intervals) we can plot a picture (see Figure 12.1), where the heights of the rectangles are proportional to the number of observations in the chosen interval ▷.

▷
A histogram.

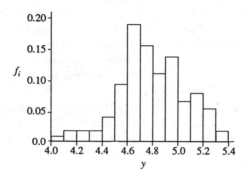

Figure 12.1
A relative frequency
diagram

It is usual to typify the values in a given interval by specifying the mid-point of the interval. It is then common to say that the height of the rectangle represents the frequency of the occurrence of the mid-point value in the measurements. Since the absolute frequency is seldom of interest, frequency is generally quoted as a *relative frequency*, f_j, obtained by dividing the number of points in the interval by the total number of points (in the present case m). Thus, to write the mean of any y_i in terms of the relative frequency, $f_{i,j}$, we simply write (dropping the i subscript for ease of writing)

$$\bar{y} = \sum_{j=1}^{m} f_j y_j, \qquad f_j = \frac{1}{m} \quad \text{with} \quad \sum_{j=1}^{m} f_j = 1,$$

where repeated values of y_j are counted as often as they occur.

It is the relative frequency which is plotted along the vertical axis in Figure 12.1, and the horizontal axis is the measured variable y. Of course, the relative frequency is given for the specified interval chosen. We can also estimate the spread in y values about the sample mean by using the sample variances, $s_y^2/(m-1)$, where s_y^2 is defined in Equation (12.5) for the present sample of m values. It may seem a bit odd to divide s_y^2 by $(m-1)$, rather than m, but the reason for this choice will become clear later.

Now suppose that we had performed this experiment, recording the m values of y, and that we had then decided to make one more measurement. Given the distribution above, we should be a little surprised if the new value did not lie close to the centre of the distribution, while recognizing that it was possible, without error in the experimental process, that it might not. If we now

▷

See, for example,
Snecedor and Cochran
(classical); Davies and
Goldsmith (classical, but
with problems drawn
from the chemical
industry); Wonnacott and
Wonnacott (classical and
Bayesian); and Lee
(Bayesian).

made a new *set* of measurements, $y_{i,(m+1)}, y_{i,(m+2)}, y_{i,(m+3)}, \ldots$ and so on, we should be very surprised if the mean from these measurements was not pretty close to the mean from the previous set of measurements. Were it not to be, then we should almost certainly start thinking about possible failures in experimental technique. However if none was to be found, the only thing that we could do would be to add the second set of points to the first, to recompute the relative frequency distribution and try again on a third set of points. The process just described is typical of aspects of modern statistical argument. On the basis of whatever is known, we try to establish the distribution of the measurements. If nothing much is known then we just have to make an informed guess. We then use the distribution to anticipate the outcome of another measurement or set of measurements. If we are not surprised by the outcomes, then we accept the initial choice that we made for the distribution. If we are surprised, then we modify the initially chosen distribution using the information gained in the trials, in the hope of constructing a distribution which will enable us to avoid surprises in future. This way of dealing with observations originates with Thomas Bayes, an eighteenth century English clergyman, and his ideas, and their subsequent development, have led to a particular way of looking at statistics which is called Bayesian. Accounts of classical and Bayesian statistics are found in several sources. ▷

12.4   The normal distribution

If repeated measurements of y for some x_i are made, we can think of the process as yielding a continuous variable $y^{(i)}$ composed of all the $y_{i,j}$ as j increases without limit. Then the relative frequency diagram could be regarded as a discrete representation of a continuous distribution function. The continuous function that is used, on the basis of experience, to model the distribution of points in this sort of case is the *normal* distribution. This is defined as

$$z = \frac{1}{\sigma\sqrt{2\pi}} e^{-((y-\mu)/\sigma)^2/2}, \qquad (12.7)$$

where the superscript on y has been dropped for ease of exposition. The constant factor multiplying the exponential is chosen to make a proper relative frequency function, in the sense that it integrates to unity. Such a constant is often called a *normalization* constant. The curve is sketched in Figure 12.2.

Now z describes the distribution of the population consisting of all possible values of y. Simply looking at the form of the function above, we should expect μ to be a measure of central tendency in the distribution and σ a measure of spread. In fact μ is the population mean and σ^2 is called the population *variance*. The (positive) square root of the variance is called the *standard deviation*. Since the normal distribution is a standard two-parameter

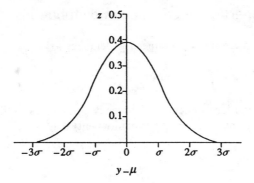

Figure 12.2
The normal distribution

function, shorthand designations have grown up for it, and it is common to denote it as: $y \sim N(\mu, \sigma^2)$, which is read as 'y is normal with mean μ and variance σ^2'.

It would be prudent to state the position of the normal distribution in this context. When we are faced with a set of experimental measurements of the kind imagined above, then it is an informed guess to treat them as if they are drawn from a normal population. On this basis, the points that are actually measured are to be regarded as a sample drawn at *random* from this population (a random sample is, roughly speaking, one that has all the characteristics of the population). With the aid of this sample we can estimate (in ways to be shown shortly) the population mean and variance. We can then use this population distribution to attempt to say something about further measurements. If what is measured surprises us, and no experimental failings can be found, then we must review the parameters at least, and possibly at length, and re-examine the form of the distribution itself. In the Bayesian approach, this process is formalized in terms of what are called prior and posterior distributions, and these are connected by a likelihood function.

However, whatever the approach adopted, there is nothing miraculous about the statistical treatment of measurements; furthermore, the statistics we obtain do not tell us that the measurements are normally distributed. It is accumulated experience that leads us to attempt a first treatment of measurements as though they were normally distributed. Doing this, we are more usually contented than surprised by subsequent measurements; but, without the subsequent measurements, any conclusions drawn on the basis of an assumption of normal distribution are simply conclusions based on this assumption. Statistical techniques enable us to make the most of the figures that we have, and to present their salient characteristics in a neat and effective way.

Before going on to discuss sampling from a normal distribution it is necessary to say a little about such ideas as mean and variance for continuous distributions.

Properties of continuous distributions

The definition of the mean generalizes to

$$\mu = \int_{-\infty}^{\infty} yf(y)dy \tag{12.8}$$

and the idea of relative frequency generalizes to the requirement (sometimes called the normalization requirement) that

$$\int_{-\infty}^{\infty} f(y)dy = 1. \tag{12.9}$$

The variance of a continuous distribution generalizes the idea of sample variance introduced above, and assumes the form

$$\sigma^2 = \int_{-\infty}^{\infty} (y - \mu)^2 f(y)dy. \tag{12.10}$$

It should be stressed that these forms are general and do not depend on $f(y)$ being the normal distribution. However, the results given in Equations (12.8), (12.9) and (12.10) are verified in the exercise below for the normal distribution, Equation (12.7).

If we think of y as having units, then dy has the same units and $f(y)$ must have the reciprocal of these units. Thus in these units $f(y)$ is a *density* and since the relative frequency must integrate to 1 it is natural to think of it as being a probability density. This accords with the standard interpretation that, for any probability density function, the probability of finding y in the interval $[a, b]$ is just given by the definite integral

$$P(a \leq y \leq b) = \int_{a}^{b} f(y)dy. \tag{12.11}$$

Extending the use of the language of probability, the mean as given in Equation (12.8) is often called the *expectation* value of y for the distribution. This idea can be extended to any function of y, say $g(y)$, so that

$$E[g(y)] = \int_{-\infty}^{\infty} g(y)f(y)dy$$

is the expectation value of $g(y)$ for the distribution $f(y)$. Expectation values can also be defined, in the obvious way, for subsets of the whole domain.

In the particular case where $g(y)$ is a power of y then such expectation values are called *moments* of the distribution, the one associated with y^k being called the kth moment. Moments can also be defined about fixed points and, in this usage, the variance is called the second moment about the mean.

Problem 12.3

Choosing $f(y)$ to be the normal distribution as in Equation (12.7), determine μ and σ^2 using Equations (12.8) and (12.10) and show that equation (12.9) is true. You will need to use the results:

$$\int_0^\infty x^{2n} e^{-ax^2}\, dx = \frac{1.3 \cdots (2n-1)}{2^{n+1}} \sqrt{\frac{\pi}{a^{2n+1}}} \quad \text{and}$$

$$\int_0^\infty x^{2n+1} e^{-ax^2}\, dx = \frac{n!}{2a^{n+1}}.$$

The fact that these are integrals over the half-range should be remembered and in the first integral, when $n = 0$, the factor multiplying the square root is just $1/2$ (see Section 11.3) and for the second integral see Problems 6.17 and 10.5.

We are now in a position to consider how we would estimate the population parameters from the sample statistics available to us.

12.5   Sampling from a distribution of measured values

Let us begin by assuming that the sample of m points that we have taken from the population of all y is a random one. In practice we cannot know this but, providing that care has been taken to remove all the sources of bias, then it is reasonable to start with this as an assumption. If subsequently we are suspicious of the results, then we must re-examine the assumption of randomness of the sample. But assuming that all is well, it can be shown that the sample mean $\bar{y}$ of a random sample chosen from a normal population is an *unbiased* estimate of the population mean. The notion of unbiasedness is intuitively clear but its technical definition is that:

$$E[\bar{y}] = \mu. \tag{12.12}$$

It can similarly be shown that the sample variance is related to the population variance as

$$E\left[\frac{s_y^2}{m-1}\right] = \sigma^2, \tag{12.13}$$

so that the unbiased estimate of σ^2 is

$$s_y^2/(m-1) \tag{12.14}$$

and this accounts for what might have seemed the rather odd choice made earlier of the definition of sample variance.

This last result means, unsurprisingly, that although a sample of only one measurement will provide an unbiased estimate of the mean, it is not sufficient to provide an estimate of the variance. The number $(m-1)$ is called the *number of degrees of freedom* in the problem. Its value is often rationalized by saying that initially we have m degrees of freedom because we have m

measurements. In order to estimate the variance, however, we must first calculate the mean and doing that removes one degree of freedom for the calculation of the variance. It is true that this rationalization often gives the right answer but actually the number of degrees of freedom arises from the mathematical process of calculating expectations and sometimes the answer is not quite as obvious as in the present case.

The results in Equations (12.12) and (12.13) above are true for a normal distribution (and indeed rather more generally). They are proved in statistics books like Snecedor and Cochran and these should be consulted for further details.

Properties of the normal distribution

The normal distribution has a most interesting property that is of the greatest generality and it is given here without proof. It is usually called the *central limit* theorem and it asserts (roughly) that if random samples of size m are chosen from *any* population then, for large enough m ($m \geq 30$ seems about right), the random sample means are distributed normally about the population mean, μ, with variance σ^2/m. This variance is called the *variance in the mean* and its square root is called the *standard error in the mean*. This is sometimes written as $\sigma_{\bar{y}}$ or denoted by SE.

In a random sample from *any* distribution, the sample mean is an unbiased estimate of the population mean and usually the sample variance can be processed in the manner given above to provide an unbiased estimate of the population variance.

Many statistical programs provide for the user to draw random samples from various kinds of distribution and the reader is recommended to try such sampling from the distributions and to prepare a frequency diagram of the sample means and to do this as a function of sample size. The theorem will not, of course, be proved by this process but we have little doubt that it will provide a convincing demonstration.

Measures of statistical confidence

We now turn to the way in which we can quantify the ideas of being surprised by, or contented with, the outcome of further measurements. Suppose for the moment that we *knew* that the population on which we were drawing was truly normal and that we knew the associated values of μ and σ. Looking at Figure 12.2 it can be seen that the value of a randomly sampled y is overwhelmingly likely to lie in a region $\pm 2\sigma$ about the mean. In fact, using Equation (12.11) with Equation (12.7) as $f(y)$ gives the probability of finding a value in the interval $[\mu - a, \mu + a]$ as:

$$P(\mu - \alpha \leq y \leq \mu + \alpha) = \frac{1}{\sigma\sqrt{2\pi}} \int_{\mu-\alpha}^{\mu+\alpha} e^{-((y-\mu)/\sigma)^2/2} \, dy. \qquad (12.15)$$

If values of α are chosen to be 1.96σ, σ, and 3σ, then the respective values for P are 0.95, 0.68 and 0.997.

In fact it is often rather easier to work in terms of a new variable, commonly called Z, which is a pure number, and defined as:

$$Z = \frac{y - \mu}{\sigma}\,, \tag{12.16}$$

so that the standardized normal distribution takes the form

$$\frac{1}{\sqrt{2\pi}} e^{-\frac{Z^2}{2}}\,,$$

with zero mean and unit variance. Thus in the shorthand notation introduced earlier we should write it as $Z \sim N(0, 1)$.

It is common also to write the probability range in Equation (12.15) in terms of $Z_\alpha \sigma$ where what is actually meant in this context is the value of Z_α that provides the required probability interval. For the 95% interval in a normal distribution, the value of α is written 0.025 to indicate that, since the probability of lying outside the interval is 0.05, that of lying in one tail is 0.025. Clearly the relation between α and interval expressed as a percentage is just $100(1 - 2\alpha)\%$. Tabulations of the normal probability associated with particular values of Z_α can be found in most statistics books.

Starting from this, we can see what can be done in the way of estimation. Let us suppose that we knew the population variance but not the mean. Given a sampled value y from the normal population, we can assert that there is a $100(1 - 2\alpha)\%$ chance that y is in the interval $\mu - Z_\alpha \sigma \leq y \leq \mu + Z_\alpha \sigma$, which rearranges to yield the corresponding interval for μ:

$$y - Z_\alpha \sigma \leq \mu \leq y + Z_\alpha \sigma\,. \tag{12.17}$$

Thus, the unknown μ lies in the known interval as specified above, with a probability $(1 - 2\alpha)$. It is usual to speak of such an interval as a $100(1 - 2\alpha)\%$ *confidence interval* and it is the custom in books on statistics for scientists and technologists to write it as $\mu = y \pm Z_\alpha \sigma$.

We know that the mean, $\bar{y}$, of a random sample is an unbiased estimate of μ for any population, and we also know that, if the sample size m of such means is suitably large, then the means are distributed normally about μ with standard error (SE) $\sigma/\sqrt{m}$. Thus we could specify the 95% confidence interval for the mean as:

$$\mu = \bar{y} \pm 1.96\sigma/\sqrt{m} \equiv \bar{y} \pm 1.96\sigma_{\bar{y}}\,.$$

Confidence limits on the mean of *any* population can be obtained for a large enough random sample from it, if we know σ. But as the latter parameter is unknown, we have to find a way of incorporating an estimate of σ into our arguments. Clearly, an estimate of σ^2 is provided by Equation (12.14) and it is sensible therefore to replace the standard error in the mean by the estimate:

$$\sigma_{\bar{y}}^2 = \frac{\sigma^2}{m} \rightarrow \frac{1}{m}\left(\frac{s_y^2}{(m-1)}\right) = s_{\bar{y}}^2\,.$$

▷

Student was the name
used by W. S. Gosset,
who worked for a famous
Dublin brewery, in order
to publish in 1908.

▷

Determined so that
$\int_{-\infty}^{\infty} z \, dq = 1$.

Unfortunately, the quantity $q = (\bar{y} - \mu)/s_{\bar{y}}$, which would be a natural generalization of Z in Equation (12.16), is not distributed normally. In fact, under reasonable assumptions on the distribution of $s_{\bar{y}}^2$, q is distributed according to Student's t distribution ▷:

$$z = C\left(1 + \frac{q^2}{v}\right)^{-\frac{v+1}{2}}$$

or $q \sim t_v$. Here C is a normalization constant ▷ so that the distribution is one of relative frequency and, since there are $m - 1$ degrees of freedom, v, we can write $q \sim t_{m-1}$. Given the definition of the exponential in Chapter 2 it is easy to see that as v increases without limit q is distributed more and more like the normal variable, Z. Essentially the t distribution is like the normal distribution but flatter and wider. We give a comparison of the standard normal distribution and the t distribution for $v = 4$ in Figure 12.3.

> **Problem 12.4**
>
> For the Student's t distribution with $v = 4$, use the results in Problems 3.2(b) and 6.3(e) to show that $\int_{-\infty}^{\infty} z \, dq = 1$ requires C to have the value 3/8.

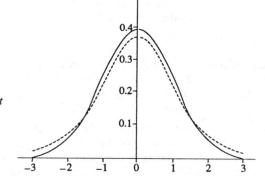

Figure 12.3
The normal (full line) and t (dashed line, $v = 4$) distributions, z, compared, with independent variables $(y - \mu)/\sigma$ and $(\bar{\mu} - \mu)/s_{\bar{y}}$, respectively

It is usual to use t to designate both the variable in the t distribution and the name of the distribution. Confusing though this may seem, the meaning is always apparent from the context.

By analogy with Equation (12.17) the confidence interval on the mean is written as $\mu = \bar{y} \pm t_{\alpha,m-1}s_{\bar{y}}$, so that the 95% confidence limits would involve $t_{0.025,m-1}$. In the case of a sample of size 21 this has the value 2.086 and by the time the sample is of size 121, it has dropped to 1.98, which is not too far from the normal distribution value of 1.96. This result provides the confidence limits on the mean for any population provided that the sample drawn from it is random.

The values $t_{\alpha,n-1}$ are tabulated in most books on statistics. The only thing to be careful of is that in some books the value of 0.025 for α used here would be quoted as 0.05 to reflect the fact that there is a 5% chance of the point lying outside the range and in either tail. Tables of this kind are called *two-tail tables*, and we just have to check carefully that we are using the appropriate tables.

In the light of the discussion above, it is now perhaps a bit easier to see why it is not easy to estimate confidence limits on r. Suppose that the population parameter appropriate to r is ρ, with a value 0.8. A sample r from such a population can then exceed this value only by 0.2 but it can be as much as 1.8 less than it. This means that the distribution of r is very skewed indeed. Of course, if there is no real linear relationship then ρ would be zero and in that special case r would be distributed in equal intervals about zero. r is thus sampled from a very odd distribution indeed and the books cited earlier should be consulted to find out more. In any event, r should be treated with great circumspection as a statistic. Unless there is some good theoretical reason for accepting a linear relationship between variables, the mere fact of $|r| \simeq 1$ should not be taken as too strong a confirmation of the existence of such a relationship.

12.6   Confidence limits on regression calculations

If we have established that a straight line relationship between our two sets of data is meaningful then, assuming that each y_i is from a normal population for each x_i and that the distribution has the same variance, whatever the value of x_i, then it can be shown that the slope b as calculated from Equation (12.3) is distributed normally about a population value β which is estimated as the expected value of b. If the population variance is σ^2 then the sampling variance of b is given by σ^2/s_x^2, where s_x^2 is given by Equation (12.5). Since σ is not known, it must be estimated and its best estimate is provided by

$$s_0^2 = \frac{1}{(n-2)} \sum_{i=1}^{n} (y_i - \hat{y}_i)^2 ,$$

so that it is closely related to U of Equation (12.1). s_0 is often called the *standard error in the estimate*. It should be emphasized here again that we are not assuming repeated measurements of y_i for any x_i nor indeed repeated measurements of x_i; we are dealing simply with the n data points as given. It is usual to speak of s_0 as being estimated with $n-2$ degrees of freedom and this is not unreasonable intuitively, because for $n=2$ the data could be fitted by a perfect straight line and there would be no deviations. In practice we need more than two points to say anything sensible.

Since we are estimating an unkown mean from a normal distribution, using an estimated sampling variance, we can expect the quantity

$$\frac{b - \beta}{s_0/s_x}$$

to be distributed according to the t distribution with $n-2$ degrees of freedom.

The confidence limits on β will be of the form:

$$\beta = b \pm t_{\alpha, n-2} \frac{s_0}{s_x} .$$

If we consider any calculated point, $\hat{y}_i$, in relation to the (unkown) true mean value $\bar{y}_i$ it turns out that the quantity $(\hat{y}_i - \bar{y}_i)/V(\hat{y}_i)$ is also distributed according to the t distribution with $n - 2$ degrees of freedom, where the sampling variance, $V^2(\hat{y}_i)$, is given by

$$s_0^2 \left(\frac{1}{n} + \frac{(x_i - \bar{x})^2}{s_x^2} \right) .$$

The confidence limits are then of the form

$$\bar{y}_i = \hat{y}_i \pm t_{\alpha, n-2} V(\hat{y}_i) , \tag{12.18}$$

demonstrating that the confidence limits become wider the further x_i moves from $\bar{x}$. Thus if we attempted an estimate outside the range of the measured values we could not be very confident in its value and certainly this is an intuitively appealing aspect of this analysis. Also, since the intercept a is the value of y when x is zero, the formula for the variance in the estimate of y provides a value for the variance in a on substituting $x_i = 0$ into Equation (12.18). Again if $x_i = 0$ is far from the mean of x then the confidence limits will have a very wide spread. The quantity s_0/s_x is called the standard error in the slope, while the quantity $V(a)$ is called the standard error in the intercept.

If x and y were independent then β would be zero; and if the true intercept were zero then $\bar{y}_0$, the true mean at $x = 0$, would be zero. It is quite usual to quote the variable values for these special cases as t values in the output of programs and to quote too the significance of these t values. A very low value of the significance of t for the slope can be taken as indicating that the regression is probably sensible and that the slope is meaningful. On the other hand a high significance value for the intercept means that the true intercept is probably zero.

If we look at the results of Example 12.1, the standard error in the slope turns out to be 1.262×10^{-7} while the standard error in the intercept is 3.092×10^{-4}. The appropriate t value for the slope is -2885.302 which has a significance value of less than $0.000\,05$ while that for the intercept is 3.0921×10^{-4} with a significance level of 0.6516. It follows, therefore, that the straight line fit is highly significant. This can also be confirmed by calculating the relevant confidence intervals.

We are now in a position to reconsider Problem 2.10 in the light of the techniques developed in this Chapter.

▷
See Barrow, p. 711.

Problem 12.5

Use the data given in Problem 2.10 for the reaction ▷

$$CH_3I + C_2H_5OH \rightarrow CH_3OC_2H_5 + HI$$

in ethanol to determine the best values of E and of A. Specify the 95% confidence limits on these values.

▷

See Raley, Rust and
Vaughan.

Problem 12.5

For the thermal gas phase decomposition of di-t-butyl peroxide

$$(CH_3)_3COOC(CH_3)_3 \longrightarrow 2(CH_3)_2CO + C_2H_6,$$

Table 12.1 gives the data for the total pressure, p, of the system (in millimetres of mercury) versus time, t, at two temperatures ▷.

Table 12.1

T/K:	427.86		420.36	
t/min	p/mm	t/min	p/mm
0	173.5	0	182.6
2	187.3	2	190.5
3	193.4	6	201.7
5	205.3	10	213.6
6	211.3	14	224.3
8	222.9	18	235.0
9	228.6	20	240.4
11	239.8	22	245.4
12	244.4	26	255.6
14	254.5	30	265.2
15	259.2	34	274.4
17	268.7	38	283.3
18	273.9	40	288.0
20	282.0	42	292.0
21	286.8	46	300.2

If the initial peroxide pressures at 427.86 K and 420.36 K are 4.2 mm and 3.1 mm, respectively, determine the first-order rate constant, k_1, for the disappearance of di-t-butyl peroxide. Calculate the 95% confidence limits on k_1.

Summary: This chapter provides an overview of the statistical tools that are necessary to deploy in the processing of experimental results. The problems associated with the fitting of a straight line expression to a given set of data are discussed at some length, and the analysis for determining appropriate assessments of the error in the intercept and slope are presented. We return to the problem of fitting data with algebraic expressions in Chapter 17; in the meantime the focus of the discussion moves from the calculus to linear algebra, the first chapter on which is concerned with the theory and use of matrices.

13 Matrices – a useful tool and a form of mathematical shorthand

Objectives

This chapter is concerned with

- matrix notation within a chemical context
- the form and properties of special kinds of matrices
- operations on matrices
- laying the foundations for vector algebra

A matrix is an *array* of elements comprising n rows and m columns, where n and m are integers greater than or equal to 1. The elements of a matrix are usually complex numbers (this includes real numbers as a special case); they can, however, be matrices themselves, or functions. The matrix notation provides a way of collecting together sets of objects with a particular kind of interrelationship. Once we have established the notation, then the properties of matrices can be explored in more detail.

A good example in the use of the matrix notation is provided by the analysis of the nature of π-bonding in planar unsaturated molecules such as a polyene, benzene, anthracene, etc. To take benzene as an example: first number the carbon atoms $1, 2, \ldots, 6$. We can then make a list of all pairs of atoms:

$$
\begin{array}{cccccc}
1-1 & 1-2 & 1-3 & 1-4 & 1-5 & 1-6 \\
2-1 & 2-2 & 2-3 & 2-4 & 2-5 & 2-6 \\
3-1 & 3-2 & 3-3 & 3-4 & 3-5 & 3-6 \\
4-1 & 4-2 & 4-3 & 4-4 & 4-5 & 4-6 \\
5-1 & 5-2 & 5-3 & 5-4 & 5-5 & 5-6 \\
6-1 & 6-2 & 6-3 & 6-4 & 6-5 & 6-6
\end{array}
$$

If we now construct an array with elements 0 or 1 to indicate the presence or absence of a π-bond between atoms i and j, respectively, then the connectivity of the carbon atoms can be represented by means of the following matrix:

$$\begin{pmatrix} 0 & 1 & 0 & 0 & 0 & 1 \\ 1 & 0 & 1 & 0 & 0 & 0 \\ 0 & 1 & 0 & 1 & 0 & 0 \\ 0 & 0 & 1 & 0 & 1 & 0 \\ 0 & 0 & 0 & 1 & 0 & 1 \\ 1 & 0 & 0 & 0 & 1 & 0 \end{pmatrix}$$

The advantage of this association of a structural formula with a matrix is that the tools of mathematics can be deployed to uncover further interesting properties and interrelationships between families of molecules. This is particularly the case in graph theory and its applications – see, for example, Gutman and Polansky. A matrix is usually designated by a printed symbol in bold-face type, for example, **A**.

The example of the matrix given above has six rows and six columns: it is referred to as a 6 by 6 matrix (usually written as a 6×6 matrix), or a square matrix of order 6. In general a matrix can be rectangular or square, as the following further examples of matrices with integer elements indicate.

$$\mathbf{A} = \begin{pmatrix} 4 & 6 & -1 \\ 3 & 0 & 2 \\ 1 & -2 & 5 \end{pmatrix} ; \ \mathbf{B} = \begin{pmatrix} 1 \\ 1 \\ 1 \end{pmatrix} ; \ \mathbf{C} = \begin{pmatrix} 3 & 1 & 2 \end{pmatrix}. \tag{13.1}$$

Here **A** has *three* rows and *three* columns and is a 3×3 matrix; **B** is a 3×1 matrix, and **C** is a 1×3 matrix.

Instead of writing matrices with specific numbers as elements, it is more convenient to generalize the notation and indicate elements by subscripted letters (very often of the same letter used to name the matrix):

$$\mathbf{A} = \begin{pmatrix} a_{11} & a_{12} & \cdots & a_{1m} \\ a_{21} & a_{22} & \cdots & a_{2m} \\ \vdots & \vdots & \ddots & \vdots \\ a_{n1} & a_{n2} & \cdots & a_{nm} \end{pmatrix} n \times m \, .$$

The first and second subscripts designate the row and column in which the element is situated: that is, $(\mathbf{A})_{ij} = a_{ij}$ is the element lying at the intersection of the ith row and the jth column of matrix **A**.

An important point to notice is that two matrices **A** and **B** are equal if and only if both have the same *dimensions* (same number of rows and columns) and also equality of all elements, such that

$$a_{ij} = b_{ij} \text{ for } \textbf{all } i, j.$$

The other properties relating to the rules for combining matrices, special forms of matrices, and operations on matrices are described in the following sections.

13.1   Rules for matrix combination

(1) If two matrices have the same dimensions then the sum and difference of the two matrices is defined as another matrix

$$C = A \pm B,$$

with the same dimensions, and elements related according to

$$(C)_{ij} = c_{ij} = a_{ij} \pm b_{ij} \text{ for all } i, j.$$

(2) A matrix may be multiplied by an ordinary number k (a *scalar*) according to the rule

$$B = kA, \text{ with}(B)_{ij} = b_{ij} = ka_{ij} \text{ for all } i,j.$$

The rule for multiplication of matrices will be introduced after an example and a problem on the first two rules have been considered.

Worked example

13.1 For the matrices D and E given below, form the sum and difference $F = D + E$ and $G = D - E$, respectively.

$$D = \begin{pmatrix} 2 & 4 \\ 0 & 1 \\ -1 & 2 \end{pmatrix}, \quad E = \begin{pmatrix} 1 & 3 \\ 2 & 1 \\ 1 & 4 \end{pmatrix}$$

Solution $$D + E = \begin{pmatrix} 3 & 7 \\ 2 & 2 \\ 0 & 6 \end{pmatrix} = F$$

$$D - E = \begin{pmatrix} 1 & 1 \\ -2 & 0 \\ -2 & -2 \end{pmatrix} = G.$$

Problem 13.1

Using the matrices defined in Example 13.1 above,

(a) show that $F + G = 2D$ and $G - F = -2E$
(b) compute $H = 3D - E$.

(3) An $n \times m$ matrix $\mathbf{A}$ and an $m \times p$ matrix $\mathbf{B}$ (notice $\mathbf{A}$ must have the same number of *columns* as $\mathbf{B}$ has *rows*) may be multiplied together to yield a matrix $\mathbf{C}$ which is an $n \times p$ matrix:

$$\underset{(n \times p)}{\mathbf{C}} = \underset{(n \times m)}{\mathbf{A}} \quad \underset{(m \times p)}{\mathbf{B}}$$

The ijth element of the product matrix $\mathbf{C}$ is defined by multiplying the ith row of $\mathbf{A}$ into the jth column of $\mathbf{B}$, element by element, and summing the products so obtained ▷:

$$(\mathbf{C})_{ij} = \sum_{k=1}^{m} (\mathbf{A})_{ik}(\mathbf{B})_{kj} \Rightarrow c_{ij}$$

$$= a_{i1}b_{1j} + a_{i2}b_{2j} + a_{i3}b_{3j} + \cdots + a_{im}b_{mj} \qquad (13.1)$$

which is written as (see Figure 13.1):

$$c_{ij} = \sum_{k=1}^{m} a_{ik}b_{kj} . \qquad (13.2)$$

As already noted, unless $\mathbf{A}$ has the same number of columns as $\mathbf{B}$ has rows, multiplication is not defined. However, even if the $n \times p$ matrix $\mathbf{AB}$ is defined, then $\mathbf{BA}$ is defined only if $n = p$. For example, with $\mathbf{B}$ and $\mathbf{C}$ as given in Equation 13.1, $\mathbf{BC}$ and $\mathbf{CB}$ are defined, and are 3×3 and 1×1, respectively – just illustrating that if both product matrices are defined then $\mathbf{BC}$ will generally have a different number of rows and columns from $\mathbf{CB}$.

When $\mathbf{A}$ and $\mathbf{B}$ are both square $n \times n$ matrices, the order of multiplication generally yields different results: that is, $\mathbf{AB}$ is a different matrix from $\mathbf{BA}$. The difference matrix, $\mathbf{AB} - \mathbf{BA}$, often written $[\mathbf{A}, \mathbf{B}]$, is called the *commutator* of $\mathbf{A}$ and $\mathbf{B}$. Matrix algebra is non-commutative, and the order of multiplication must always be observed.

There is no general way of defining matrix division although, for certain kinds of square matrix, it is possible to define a related matrix which plays the role of a divisor. We return to this problem of the construction of this related matrix after an excursion into determinants in the next chapter.

▷
k is a counting, or dummy, index and no other significance should be attached to its name.

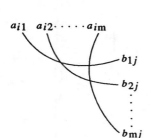

Figure 13.1
Construct to illustrate the formation of c_{ij} from the ith row of A and the jth column of B

Worked example

13.2 For the matrices

$$\mathbf{X} = \begin{pmatrix} 1 & 5 \\ 2 & 7 \\ 3 & 4 \end{pmatrix} \quad \mathbf{Y} = \begin{pmatrix} 8 & 4 & 3 & 1 \\ 2 & 5 & 8 & 6 \end{pmatrix}$$

(a) determine whether $\mathbf{XY}$ and $\mathbf{YX}$ are defined;

(b) where the product is defined, compute the product matrix $\mathbf{Z}$.

Solution   (a) $\mathbf{X}$ is a 3×2 and $\mathbf{Y}$ a 2×4 matrix. Hence only the 3×4 product matrix $\mathbf{Z} = \mathbf{XY}$ is defined.

(b)

$$\mathbf{XY} = \begin{pmatrix} 1 & 5 \\ 2 & 7 \\ 3 & 4 \end{pmatrix} \begin{pmatrix} 8 & 4 & 3 & 1 \\ 2 & 5 & 8 & 6 \end{pmatrix}$$

$$= \begin{pmatrix} 1 \cdot 8 + 5 \cdot 2 & 1 \cdot 4 + 5 \cdot 5 & 1 \cdot 3 + 5 \cdot 8 & 1 \cdot 1 + 5 \cdot 6 \\ 2 \cdot 8 + 7 \cdot 2 & 2 \cdot 4 + 7 \cdot 5 & 2 \cdot 3 + 7 \cdot 8 & 2 \cdot 1 + 7 \cdot 6 \\ 3 \cdot 8 + 4 \cdot 2 & 3 \cdot 4 + 4 \cdot 5 & 3 \cdot 3 + 4 \cdot 8 & 3 \cdot 1 + 4 \cdot 6 \end{pmatrix}$$

$$= \begin{pmatrix} 8 + 10 & 4 + 25 & 3 + 40 & 1 + 30 \\ 16 + 14 & 8 + 35 & 6 + 56 & 2 + 42 \\ 24 + 8 & 12 + 20 & 9 + 32 & 3 + 24 \end{pmatrix}$$

$$\Rightarrow \mathbf{Z} = \begin{pmatrix} 18 & 29 & 43 & 31 \\ 30 & 43 & 62 & 44 \\ 32 & 32 & 41 & 27 \end{pmatrix}.$$

> **Problem 13.2**
>
> Using the matrices given in Equation (13.1) and Example 13.1, determine, where defined, the resultant matrices:
>
> (a) $\mathbf{AD}$ (b) $\mathbf{DA}$ (c) $(\mathbf{CA})\mathbf{D}$ (d) $\mathbf{C(AD)}$ (e) $\mathbf{CE}$ (f) $\mathbf{AE}$ (g) $\mathbf{DE}$ (h) $\mathbf{AC}$ (i) $\mathbf{CA}$ (j) $\mathbf{A(D + E)}$ (k) $\mathbf{AD + AE}$.

▷
An operation like multiplication or addition that combines two objects (here matrices).

The results of Problem 13.2 illustrate examples of two other important properties of matrices when binary operations ▷ are defined: *associativity* when multiplication is defined (the equality of $(\mathbf{AB})\mathbf{C}$ and $\mathbf{A(BC)}$), and *distributivity* (the equality of $\mathbf{A(B + C)}$ and $\mathbf{AB + AC}$).

13.2   Special forms of matrices and operations on matrices

This section is concerned with forms of matrices that arise sufficiently frequently to make it worthwhile to give them special names. It is also timely to describe how some important operations on matrices are defined in order to prepare the way for an introductory discussion of group theory towards the end of this chapter, and also further discussions on the use of matrices in later chapters.

The null matrix

The *null* matrix is an $n \times m$ matrix all of whose elements are null (zero). It is often written as $\mathbf{O}_{n \times m}$, for a *rectangular* matrix ($m \neq n$), and $\mathbf{O}_n$ for a *square* matrix ($m = n$). Often the subscripts are omitted since they can be inferred from the context. Given any other matrix $\mathbf{X}$ which is $m \times n$ then

$$\mathbf{O}_{n \times m} \mathbf{X} = \mathbf{O}_n, \quad \mathbf{X} \mathbf{O}_{n \times m} = \mathbf{O}_m .$$

Results of this kind are easy to prove using the algebraic result in Equation (13.3); consider, for example, the ijth element of $\mathbf{OX}$:

$$(\mathbf{O}_{n \times m} \mathbf{X})_{ij} = \sum_k o_{ik} x_{kj}, \quad \text{for } i,j = 1, 2, \ldots, n$$
$$= 0 = o_{ij} = (\mathbf{O}_n)_{ij} .$$

Since this result holds for all i, j indicated above, rule (1) enables us to write $\mathbf{O}_{n \times m} \mathbf{X} = \mathbf{O}_n$.

The unit matrix

The *unit* matrix is a square, $n \times n$ matrix, denoted $\mathbf{E}_n$ (or just $\mathbf{E}$) such that it has unit elements along its diagonal and zero elements elsewhere. Thus

$$\mathbf{E}_3 = \begin{pmatrix} 1 & 0 & 0 \\ 0 & 1 & 0 \\ 0 & 0 & 1 \end{pmatrix}$$

If $\mathbf{X}$ is $n \times p$ then $\mathbf{E}_n \mathbf{X} = \mathbf{X}$ and $\mathbf{X} \mathbf{E}_p = \mathbf{X}$. The elements of $\mathbf{E}$ are usually denoted as δ_{ij} (and not e_{ij}), where δ_{ij} is the *Kronecker delta* defined as follows:

$$\delta_{ij} = 1 \quad \text{if} \quad i = j; \quad \delta_{ij} = 0 \quad \text{if} \quad i \neq j$$

so $(\mathbf{E})_{ij} = \delta_{ij}$. The unit matrix $\mathbf{E}$ is sometimes denoted by $\mathbf{I}$ or $\mathbf{1}$.

> **Problem 13.3**
>
> For the matrices $\mathbf{X}_{m \times p}$, $\mathbf{E}_m$, use Equation (13.3) to prove that $\mathbf{E}_m \mathbf{X} = \mathbf{X}$.

Symmetric matrices

A square matrix $\mathbf{A}$ with elements having the property $a_{ij} = a_{ji}$ is said to be a *symmetric* matrix: for example,

$$\mathbf{A} = \begin{pmatrix} 1 & 2 & 3 \\ 2 & 4 & 5 \\ 3 & 5 & 6 \end{pmatrix} \text{ is symmetric.} \tag{13.4}$$

The transpose of a matrix

If the operation of interchanging rows and columns of an $n \times m$ matrix $\mathbf{B}$ is carried out, then the *transpose* of $\mathbf{B}$, written $\mathbf{B}^T$, is obtained according to

$$\left(\mathbf{B}^T\right)_{ij} = b_{ji}.$$

Thus the first row of $\mathbf{B}$ forms the first column of $\mathbf{B}^T$ and so on, and $\mathbf{B}^T$ is an $m \times n$ matrix. The symmetric matrix given in Equation (13.4) is an example of a matrix for which $\mathbf{A}^T = \mathbf{A}$.

Worked example

13.4   For $\mathbf{B} = \begin{pmatrix} 1 & 2 \\ 3 & 4 \\ 5 & 6 \end{pmatrix}$, give $\mathbf{B}^T$.

Solution   The transpose is given by $\mathbf{B}^T = \begin{pmatrix} 1 & 3 & 5 \\ 2 & 4 & 6 \end{pmatrix}$.

Problem 13.4

Write down the transpose of each of the matrices given in Equation (13.1) ($\mathbf{A}, \mathbf{B}, \mathbf{C}$), and Example 13.1 ($\mathbf{D}, \mathbf{E}$).

Problem 13.5

(a) Show that for any $n \times m$ matrix $\mathbf{B}$, both $\mathbf{B}^T\mathbf{B}$ and $\mathbf{B}\mathbf{B}^T$ are defined, and give the dimension of the product matrix in each case.

(b) Form the two possible products of a matrix and its transpose, using first the matrix $\mathbf{D}$ given in Example 13.1 and then $\mathbf{C}$ from Equation (13.1).

It is apparent from the results of Problem 13.5 that both $\mathbf{B}^T\mathbf{B}$ and $\mathbf{B}\mathbf{B}^T$ are symmetric matrices: a consequence of that fact is that $\left(\mathbf{B}^T\right)^T = \mathbf{B}$ and that, generally,

$$\left(\mathbf{AB}\right)^T = \mathbf{B}^T\mathbf{A}^T. \tag{13.5}$$

Problem 13.6

(a) Choose $\mathbf{A} = \mathbf{B}^T$, and substitute in Equation (13.5) to verify that, in general, $\left(\mathbf{B}^T\mathbf{B}\right)$ is symmetric.

> (b) Demonstrate that Equation (13.5) is true when $\mathbf{A}$ and $\mathbf{B}$ are taken as the matrices $\mathbf{A}$ and $\mathbf{D}$ from Equation (13.1) and Example 13.1, respectively.
>
> (c) Let $\mathbf{C} = \mathbf{AB}$, and use the definition of matrix multiplication, given in Equation (13.3), to express the ijth element of $\mathbf{C}^T$ ▷ in terms of the elements of $\mathbf{A}^T$ and $\mathbf{B}^T$ in order to prove the result given in Equation (13.5).

▷
Remember that $(\mathbf{C}^T)_{ij} = c_{ji}$.

The trace of a matrix

▷
Reminder: the index m used for labelling the diagonal elements of $\mathbf{A}$ is dummy.

If $\mathbf{A}$ is a square $n \times n$ matrix, then its *trace* (written 'tr') is defined as the sum ▷ of its diagonal elements:

$$\operatorname{tr}\mathbf{A} = \sum_{m=1}^{n} = a_{mm}. \tag{13.6}$$

Thus $\operatorname{tr}\mathbf{A}$ from Equation (13.3) above is $1 + 4 + 6 = 11$. Obviously $\operatorname{tr}\mathbf{A} = \operatorname{tr}\mathbf{A}^T$, since the trace depends only upon the diagonal elements of $\mathbf{A}$.

▷
A result which also holds if $\mathbf{A}$ is $(m \times n)$ and $\mathbf{B}$ is $(n \times m)$.

Problem 13.7

If $\mathbf{A}$ and $\mathbf{B}$ are square $m \times m$ matrices then $\operatorname{tr}(\mathbf{AB}) = \sum_{p=1}^{m}(\mathbf{AB})_{pp}$ ▷.

(a) Use Equation (13.3) with $i = j = p$ to write the diagonal element of the product matrix $\mathbf{C} = \mathbf{AB}$ in terms of the elements of the pth row of $\mathbf{A}$ and the pth column of $\mathbf{B}$.

(b) Use the commutativity of the elements of matrices, and perform the summation over the index p to show that $\operatorname{tr}(\mathbf{AB}) = \sum_{k=1}^{m}(\mathbf{BA})_{kk}$.

Thus prove that $\operatorname{tr}(\mathbf{AB}) = \operatorname{tr}(\mathbf{BA})$.

(c) By substituting $\mathbf{B}^T$ for $\mathbf{A}$, or otherwise, prove that $\operatorname{tr}(\mathbf{B}^T\mathbf{B}) = \operatorname{tr}(\mathbf{BB}^T)$.

(d) By substituting $\mathbf{X}$ for $\mathbf{AB}$ and then $\mathbf{Y}$ for $\mathbf{CA}$, prove that $\operatorname{tr}(\mathbf{ABC}) = \operatorname{tr}(\mathbf{CAB}) = \operatorname{tr}(\mathbf{BCA})$.

The complex conjugate of a matrix

The complex conjugate of a matrix $\mathbf{A}$ is written as $\mathbf{A}^*$, and is defined such that $(\mathbf{A}^*)_{ij} = a_{ij}^*$; and, if all the elements of $\mathbf{A}$ are real, then $\mathbf{A}^* = \mathbf{A}$.

The adjoint of a matrix

▷

Some authors, for example Stephenson, define the adjoint matrix to be the matrix of cofactors (see Chapter 14).

The transpose of the complex conjugate of a matrix yields its adjoint, usually written as $\mathbf{A}^\dagger$, and is defined ▷ such that

$$\mathbf{A}^\dagger = (\mathbf{A}^*)^T = (\mathbf{A}^T)^*$$

or, in terms of elements, $(\mathbf{A}^\dagger)_{ij} = a_{ji}^*$. If $\mathbf{A}^* = \mathbf{A}$ (a real matrix) then $\mathbf{A}^\dagger = \mathbf{A}^T$.

> **Problem 13.8**
>
> If for two matrices $\mathbf{A}$ and $\mathbf{B}$, the product $\mathbf{AB}$ is defined, use the procedure suggested in part (c) of Problem 13.6 to demonstrate that
>
> (a) $(\mathbf{AB})^* = \mathbf{A}^*\mathbf{B}^*$, using $*$ instead of T
> (b) $(\mathbf{AB})^\dagger = \mathbf{B}^\dagger\mathbf{A}^\dagger$, using $\dagger$ instead of T.

Hermitian matrices

If $\mathbf{A}$ is a square matrix such that $\mathbf{A} = \mathbf{A}^\dagger$, then $\mathbf{A}$ is said to be an *hermitian* matrix. A real hermitian matrix is a symmetric matrix.

Orthogonal matrices

The 3×3 square matrix

$$\mathbf{R} = \begin{pmatrix} \cos\theta & -\sin\theta & 0 \\ \sin\theta & \cos\theta & 0 \\ 0 & 0 & 1 \end{pmatrix}$$

has the property that $\mathbf{R}^T\mathbf{R} = \mathbf{R}\mathbf{R}^T = \mathbf{E}$, which characterizes an *orthogonal* matrix. $\mathbf{E}$ is the unit matrix (here 3×3).

Unitary matrices

The 2×2 matrix

$$\mathbf{U} = \begin{pmatrix} x & -y^* \\ y & x^* \end{pmatrix}$$

▷

$\mathbf{U}^\dagger\mathbf{U}$ and $\mathbf{U}\mathbf{U}^\dagger$, respectively.

with complex elements such that $xx^* + yy^* = 1$, is an example of a *unitary* matrix with the property $\mathbf{U}^\dagger\mathbf{U} = \mathbf{U}\mathbf{U}^\dagger = \mathbf{E}_2$. Any square matrix $\mathbf{U}$ which yields the unit matrix when pre- or post-multiplied by its conjugate transpose (adjoint) ▷ is a unitary matrix.

Problem 13.9

The matrix **A** has the form

$$\mathbf{A} = \begin{pmatrix} 2 & 3+i \\ 3-i & 1 \end{pmatrix}.$$

(a) Show that it is hermitian.
(b) If **x** is a 2×1 matrix $\triangleright$, with x_1 and x_2 arbitrary complex numbers of the form $a_1 + ib_1$ and $a_2 + ib_2$, respectively, show that $\mathbf{x}^\dagger \mathbf{A} \mathbf{x}$ is a real number, and give its form in terms of the real and imaginary parts of x_1 and x_2.

$\triangleright$
Such matrices are often termed column matrices.

Problem 13.10

Given the matrices

$$\mathbf{E}_2 = \begin{pmatrix} 1 & 0 \\ 0 & 1 \end{pmatrix}, \quad \mathbf{J} = \begin{pmatrix} 0 & 1 \\ -1 & 0 \end{pmatrix}, \quad \mathbf{Z} = \begin{pmatrix} r\cos\theta & r\sin\theta \\ -r\sin\theta & r\cos\theta \end{pmatrix}$$

(a) show that
   (i) $\mathbf{J}\mathbf{J} = -\mathbf{E}_2$ and $\mathbf{Z} = r(\cos\theta\,\mathbf{E}_2 + \sin\theta\,\mathbf{J})$,
   (ii) $\mathbf{Z}^T = r(\cos\theta\,\mathbf{E}_2 - \sin\theta\,\mathbf{J})$ (iii) $\mathbf{Z}\mathbf{Z}^T = r^2\mathbf{E}_2$;
(b) Using the following definition of the exponential of a 2×2 matrix $\triangleright$,

$$e^{\mathbf{X}} = \mathbf{E}_2 + \mathbf{X} + \frac{1}{2!}\mathbf{X}^2 + \frac{1}{3!}\mathbf{X}^3 + \cdots,$$

where **X** is a real square 2×2 matrix, show that $\triangleright$
   (i) $re^{\theta\mathbf{J}} = \mathbf{Z}$, (ii) $e^{\mathbf{E}_2} = e\mathbf{E}_2 = \begin{pmatrix} e & 0 \\ 0 & e \end{pmatrix}.$

$\triangleright$
The exponential function here involves the mapping of a 2×2 matrix to a 2×2 matrix.
$\triangleright$
See Section 8.3 for the expansions of $\cos\theta$ and $\sin\theta$.

13.3   Isomorphisms involving matrices

The last problem illustrates several important mathematical ideas. First, it is possible, as seen here, to define a domain of square matrices for the exponential function; it is also possible to develop power series expansions of matrices which are analogues of $(1+x)^n$, where n is rational, and other functions of matrices. Second, the problem has more than a superficial likeness to the algebra of complex numbers. In particular, if the numbers 1 and i are taken to stand in a 1:1 correspondence with the matrices $\mathbf{E}_2$ and **J**, then any real number, x, or imaginary number yi, stands in a 1:1 correspondence with

$x\mathbf{E}_2$ and $y\mathbf{J}$, respectively. Thus any element of the set $\mathbb{C}$ of complex numbers displays a 1:1 association with the matrix $\mathbf{Z} = x\mathbf{E}_2 + y\mathbf{J}$. What we have here is a prescription for defining a function which maps from the domain $\mathbb{C}$ to a *subset* of 2×2 matrices of the form

$$\mathbf{Z} = x\mathbf{E}_2 + y\mathbf{J} = \begin{pmatrix} x & y \\ -y & x \end{pmatrix}. \tag{13.6}$$

▷

$z_{11} = z_{22} = x$ and
$z_{12} = -z_{21} = y$.

It is clear, moreover, that $\mathbf{Z}$ can only lie in a subset of 2×2 matrices, because a general matrix has four independent elements $a_{11}, a_{12}, a_{21}, a_{22}$ and $\mathbf{Z}$ has only two ▷.

This kind of function is usually called an *isomorphism*. Functions of this form occur widely in chemistry and are not always recognized as such: for example, in the opening paragraph of this chapter, such a function is presented where the mapping is from a member of the set of planar hydrocarbons to a subset of square matrices with elements zero and one. Another isomorphism is in the mapping of marks on an X-ray diffraction pattern to the ordered triple (h, k, l) in order to index which planes of atoms are responsible for producing the particular mark on the X-ray film. The mapping of a chemical name to an empirical formula does not, however, define an isomorphism as is seen, for example, with 3-methylbuta-1,2-diene, 3-methylbut-1-yne and twenty-four other species (including two polymers with the same empirical formula) ▷ which map to C_5H_8.

▷

See the Twelfth
Collective Index (1987–
1991) of
Chemical Abstracts.

The important point about recognizing the presence of an isomorphism within a chemical context is that, very often, it is appropriate to bring the full force of the mathematics associated with the abstract subject of isomorphisms to provide a framework for understanding the chemistry. An especially good example is seen in the way the symmetry of certain kinds of organic molecules can be utilized to provide information about the reactions they can undertake: in particular, whether a ring-opening or ring-closing step occurs under the influence of heat or light ▷. In other applications, symmetry arguments are used to determine either whether electronic or vibrational transitions are allowed, or how molecular orbitals are constructed from the constituent atomic orbitals associated with chemically equivalent atoms.

▷

Woodward and
Hoffmann.

Some properties of groups

As far as the last two examples are concerned, the key step in the understanding of these applications involves using the isomorphism between square matrices and the transformations of points in the molecule – each of which leaves the molecule in a state indistinguishable from the initial configuration. Such transformations are termed *symmetry operations*. The set of all symmetry operations for a molecule forms the point group for the molecule, and the important associated subject of *symmetry theory* is given a brief introduction in Atkins ▷ as far as electronic structure applications are concerned; other texts also provide a useful source for a wider range of applications, including vibrational spectroscopy ▷.

▷

Chapter 15.
▷
See, for example,
Bishop.

The key mathematical feature in the chemical applications of symmetry theory relates to the mathematical notion of a *group*. A group is nothing more than a set of objects (for example, numbers or matrices) which are combined under a binary operation $\triangleright$ to yield a member of the set; also there must be certain objects within the set that have specific properties under the binary operation. Let us for the moment consider the combination of the elements of the set $\{1, -1, i, -i\}$ under the binary operations of addition and then multiplication. Clearly, under addition $1 + 1 = 2$ (which is not a member of the set), and addition is therefore not an appropriate operation on the elements of this set $\triangleright$. On the other hand, under multiplication the set is closed: that is, for any two elements x, y in the set xy is also a member of the set. In order for the set to form a group, several other properties of the elements need to be tested. First, there has to be an *identity* element e such that $ex = x$ for all x in the set; second, x must possess an *inverse* – usually given the symbol x^{-1} – such that $xx^{-1} = x^{-1}x = e$; third, the rule of associativity must apply, which requires that when there are two or more binary combinations of elements, the choice of which binary combination to take first does not matter. For example, with the product xyz, the same result must obtain whether the product xy is combined with z, or x is combined with the product yz. Expressed more succinctly, this means that $(xy)z = x(yz)$ $\triangleright$.

For the given set under the operation of multiplication, we can make the following observations by direct combination of the elements:

$\triangleright$
See the comments after Problem 13.2.

$\triangleright$
Neither is subtraction.

$\triangleright$
This property of associativity is also reflected in the way matrices are defined and combined.

Element	Property
1	Identity and self-inverse
-1	Self-inverse
i	Inverse of $-i$
$-i$	Inverse of i

The set is closed, and the law of associativity holds. The set therefore forms a group under multiplication.

Problem 13.11

Show that

(a) the set of integers (including zero) forms a group under addition, but not under multiplication or subtraction. Where the set does not form a group, give the group properties that are not observed.
(b) the set of real numbers (excluding zero) forms a group under multiplication.

Problem 13.12

As seen in Equation (13.6), an isomorphism can be established between the set of real numbers (excluding zero) and matrices of the form

$$\mathbf{x} = x\begin{pmatrix} 1 & 0 \\ 0 & 1 \end{pmatrix} = \begin{pmatrix} x & 0 \\ 0 & x \end{pmatrix}.$$

Since the set of real numbers (excluding zero) forms a group under multiplication (Problem 13.11), we expect that corresponding sets of matrices, with elements $\mathbf{x}$, also form a group under multiplication.

Show that the set of matrices $\mathbf{x}$ forms a group under multiplication by demonstrating that

(a) the identity element e is $\mathbf{E}_2$

(b) the inverse of $\mathbf{x}$ is $\mathbf{x}^{-1} = \dfrac{1}{x}\mathbf{E}_2$

(c) the set is closed under multiplication.

Problem 13.13

The set of matrices $\mathbf{Z}$, representing the complex numbers (excluding zero) under the binary operation of multiplication, contains elements of the form

$$\mathbf{Z}_i = x_i\begin{pmatrix} 1 & 0 \\ 0 & 1 \end{pmatrix} + y_i\begin{pmatrix} 0 & 1 \\ -1 & 0 \end{pmatrix} = \begin{pmatrix} x_i & 0 \\ 0 & x_i \end{pmatrix} + \begin{pmatrix} 0 & y_i \\ -y_i & 0 \end{pmatrix}$$

$$= \begin{pmatrix} x_i & y_i \\ -y_i & x_i \end{pmatrix} = x_i\mathbf{E}_2 + y_i\mathbf{J}.$$

Demonstrate that

(a) the identity element is $\mathbf{E}_2$

(b) $\mathbf{Z}_i\mathbf{Z}_i^T = (x_i^2 + y_i^2)\mathbf{E}_2$, and hence deduce that

$$\mathbf{Z}_i^{-1} = \frac{1}{x_i^2 + y_i^2}(x_i\mathbf{E}_2 - y_i\mathbf{J})$$

(c) the set of $\mathbf{Z}$ matrices is closed under multiplication, by showing that the product of two matrices in the set, say

$$\mathbf{Z}_1 = \begin{pmatrix} x_1 & y_1 \\ -y_1 & x_1 \end{pmatrix} \text{ and } \mathbf{Z}_2 = \begin{pmatrix} x_2 & y_2 \\ -y_2 & x_2 \end{pmatrix}$$

is also an element in the set.

Group representations

As matrix multiplication displays the associative property, it should now be clear that the set of 2×2 **Z** matrices (excluding the null matrix $\mathbf{O_2}$) forms a group. We see again, in the problem above, that an isomorphism links the two sets of objects – in this case, the set of complex numbers (excluding zero) and the set of **Z** matrices (excluding $\mathbf{O_2}$). Because this isomorphism preserves the group property, the set of matrices is said to form a *representation* of the group of complex numbers under multiplication.

The notion of a group representation in terms of matrices is the most important aspect of group theory in chemistry. As already noted previously, the set of symmetry operations for a molecule forms a group; and, as we shall see, such operations are isomorphically linked to sets of square matrices ▷. The following discussion should help to make the nature of this isomorphism clearer, and further reading of the references already cited should clarify some of the points in the authors' use of more extensive examples of a chemical nature.

▷
The particular set of representation matrices depends upon the problem of interest.

The symmetry properties of ozone – a chemical example

In the V-shaped molecule ozone, O_3, all the atoms lie in a plane, with an $O-\widehat{O}-O$ angle of 155°. Consider the problem of exploring the constraints of symmetry in the discussion of the π-bonding in this triatomic species. It is convenient to envisage the molecule with atoms, labelled O_a, O_b, O_c, situated in the yz plane at $(0,0,0)$, $(0, -R\sin(\phi/2), R\cos(\phi/2))$, $(0, R\sin(\phi/2), R\cos(\phi/2))$, respectively, where R is the O–O bond length and ϕ is the interbond angle (Figure 13.2); the three π_x atomic orbitals, perpendicular to the yz plane containing the nuclei, are designated by p_a, p_b and p_c, respectively.

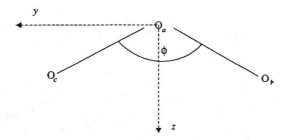

Figure 13.2
The choice of axes for O_3: the x-axis is directed out of the paper to complete a right-handed axis system

Table 13.1

Operation description	Symbol	Transformation of atomic orbitals		
Leaving the framework unmoved (identity)	E	$p_a \rightarrow p_a,$	$p_b \rightarrow p_b,$	$p_c \rightarrow p_c$
Rotation about the z-axis by 180°	$C_2(z)$	$p_a \rightarrow -p_a,$	$p_b \rightarrow -p_c,$	$p_c \rightarrow -p_b$
Reflection in the yz plane	$\sigma(yz)$	$p_a \rightarrow -p_a,$	$p_b \rightarrow -p_b,$	$p_c \rightarrow -p_c$
Reflection in the xz plane	$\sigma(xz)$	$p_a \rightarrow p_a,$	$p_b \rightarrow p_c,$	$p_c \rightarrow p_b$

The symmetry operations associated with this geometrical shape lead to the transformations on the three π-type atomic orbitals shown in Table 13.1. Notice that although the pairs of symmetry operations E, $\sigma(xz)$ and $\sigma(yz)$, $C_2(z)$ effect the same rearrangements of the points associated with the nuclear positions, they correspond to different symmetry operations insofar as the latter operation in each pair transforms a positive lobe of a π-orbital into a negative lobe – either on the same or on an equivalent centre.

For each operation, the orbitals p_a, p_b, p_c are transformed into another sequence, with a possible change in sign. The best way of handling this transformation is to collect the orbitals into a 1×3 (row) matrix and write down the row matrix of the transformed orbitals; it is a straightforward procedure to generate an (orthogonal) matrix representing each operation as follows:

$$(p_a \quad p_b \quad p_c) \xrightarrow{E} (p_a \quad p_b \quad p_c) = (p_a \quad p_b \quad p_c) \begin{pmatrix} 1 & 0 & 0 \\ 0 & 1 & 0 \\ 0 & 0 & 1 \end{pmatrix},$$

$$(p_a \quad p_b \quad p_c) \xrightarrow{C_2(z)} (-p_a \quad -p_c \quad -p_b)$$
$$= (p_a \quad p_b \quad p_c) \begin{pmatrix} -1 & 0 & 0 \\ 0 & 0 & -1 \\ 0 & -1 & 0 \end{pmatrix},$$

$$(p_a \quad p_b \quad p_c) \xrightarrow{\sigma(yz)} (-p_a \quad -p_b \quad -p_c)$$
$$= (p_a \quad p_b \quad p_c) \begin{pmatrix} -1 & 0 & 0 \\ 0 & -1 & 0 \\ 0 & 0 & -1 \end{pmatrix},$$

$$(p_a \quad p_b \quad p_c) \xrightarrow{\sigma(xz)} (p_a \quad p_c \quad p_b) = (p_a \quad p_b \quad p_c) \begin{pmatrix} 1 & 0 & 0 \\ 0 & 0 & 1 \\ 0 & 1 & 0 \end{pmatrix}.$$

If a general symmetry operation is designated by $\hat{R}$, then each of the above matrix transformations may be written more succinctly as $\hat{R}\mathbf{a} = \mathbf{a}\mathbf{D}(\hat{R})$, where $\mathbf{a}$ is the row matrix $(p_a \, p_b \, p_c)$. The most important observation here is that if two symmetry operations have the property $\hat{P}\hat{R} = \hat{S}$ then the matrices representing the respective operations have the analogous property $\mathbf{D}(\hat{P})\mathbf{D}(\hat{R}) = \mathbf{D}(\hat{S})$; thus there is an isomorphism between the set of symmetry operations and the set of representation matrices. However, as the symmetry operations form a group, we conclude that the set of matrix representatives also forms a group. Furthermore, it is the trace (or *character*) of each of the representation matrices ($3, -1, -3, 1$, respectively) which is used in elucidating the form of each of the molecular orbitals that can be constructed from the three atomic orbitals ▷.

▷
See Atkins, Chapter 15.

The isomorphism here therefore provides a mapping from a set of symmetry operations to a set of matrices – thus enabling us to carry out symmetry operations in terms of matrix algebra. The representation matrices are of order three because three atomic orbitals are used as building blocks for describing the π-bonding in O_3. On the other hand, in the analysis of the possible motions of the oxygen nuclei in O_3, the matrices generated by the nuclear displacement coordinates (three for each nucleus) are of order nine, thus illustrating that the dimension of the representation matrices varies according to the nature of the problem.

$\triangleright$

Notice that sets of atomic orbitals transform as row matrices, whilst sets of nuclear position coordinates transform as column matrices – see, for example, Atkins, Section 15.4.

Problem 13.14

Determine the representation matrices, and their characters, for the symmetry operations E, $C_2(z)$, $\sigma(xz)$ and $\sigma(yz)$ associated with O_3 that describe the transformation properties of the oxygen

(a) $2s$ atomic orbitals
(b) nuclear position coordinates (x_i, y_i, z_i), where $i = a, b, c$.

Hint for (b): the nuclear coordinates are collected into *column* rather than *row* matrices $\triangleright$. Hence, for example, under $C_2(z)$:

$$
\begin{pmatrix} x_a \\ y_a \\ z_a \\ x_b \\ y_b \\ z_b \\ x_c \\ y_c \\ z_c \end{pmatrix}
\xrightarrow{C_2(z)}
\begin{pmatrix} -x_a \\ -y_a \\ z_a \\ -x_c \\ -y_c \\ z_c \\ -x_b \\ -y_b \\ z_b \end{pmatrix}
$$

$$
= \begin{pmatrix}
-1 & 0 & 0 & 0 & 0 & 0 & 0 & 0 & 0 \\
0 & -1 & 0 & 0 & 0 & 0 & 0 & 0 & 0 \\
0 & 0 & 1 & 0 & 0 & 0 & 0 & 0 & 0 \\
0 & 0 & 0 & 0 & 0 & 0 & -1 & 0 & 0 \\
0 & 0 & 0 & 0 & 0 & 0 & 0 & -1 & 0 \\
0 & 0 & 0 & 0 & 0 & 0 & 0 & 0 & 1 \\
0 & 0 & 0 & -1 & 0 & 0 & 0 & 0 & 0 \\
0 & 0 & 0 & 0 & -1 & 0 & 0 & 0 & 0 \\
0 & 0 & 0 & 0 & 0 & 1 & 0 & 0 & 0
\end{pmatrix}
\begin{pmatrix} x_a \\ y_a \\ z_a \\ x_b \\ y_b \\ z_b \\ x_c \\ y_c \\ z_c \end{pmatrix}
$$

A much more compact notation can be introduced, however, by partitioning the 9×9 matrix into 3×3 matrices:

$$\begin{pmatrix} \mathbf{x}_a \\ \mathbf{x}_b \\ \mathbf{x}_c \end{pmatrix} \xrightarrow{C_2(z)} \begin{pmatrix} \mathbf{X}(C_2) & \mathbf{O}_3 & \mathbf{O}_3 \\ \mathbf{O}_3 & \mathbf{O}_3 & \mathbf{X}(C_2) \\ \mathbf{O}_3 & \mathbf{X}(C_2) & \mathbf{O}_3 \end{pmatrix} \begin{pmatrix} \mathbf{x}_a \\ \mathbf{x}_b \\ \mathbf{x}_c \end{pmatrix},$$

where $\mathbf{O}_3$ is the 3×3 null matrix,

$$\mathbf{X}(C_2) = \begin{pmatrix} -1 & 0 & 0 \\ 0 & -1 & 0 \\ 0 & 0 & 1 \end{pmatrix} \text{ and } x_i = \begin{pmatrix} x_i \\ y_i \\ z_i \end{pmatrix}$$

Isomorphisms between groups

The number of elements in a group determines its *order* and, for the two finite groups considered so far, their orders are four. Not all groups of chemical interest, however, are finite: thus, for example, a linear molecule like HCl has infinitely many rotations about the internuclear axis, as a rotation by *any* angle is a symmetry operation; also, any crystalline solid is characterized by an infinite number of symmetry operations associated with the translation of the contents of the unit cell.

In reconsidering the two groups of order four, the interesting question arises as to whether there is an isomorphism between them, with an associated 1:1 mapping between their two sets of elements. If an isomorphism does exist then these two groups are essentially the same: the sort of sameness that is observed when we ascribe the numeral '2' to two oranges, two people or any pair of similar objects. In fact, it is a simple matter to verify that there is no isomorphic relation between the two groups of order four considered here ▷. In the group generated by $\pm 1, \pm i$ under multiplication, the elements i and $-i$ are inverses of each other; however in the group of symmetry operations for ozone, all elements are self-inverses. It turns out that there are at least two other groups isomorphically linked with the group of ozone – one of which is the group characterizing the shape of *trans*-buta-1,3-diene. Further details of how the powerful tools of matrix algebra, and the theory of group representations, are deployed in a chemical context are found in texts on group theory ▷.

▷

There is an isomorphism, however, between $\{\pm 1, \pm i\}$ and the point groups S_4 or C_4.

▷

See, for example, Bishop or Vincent.

Summary: This completes the first chapter on the theory and use of matrices. In subsequent chapters we use matrices in the study of determinants, and also in applications of vector algebra. The range and nature of the chemical applications of matrix algebra is indeed extensive (especially in quantum chemistry and spectroscopy), and it is therefore appropriate that the remainder of this text is devoted to the further study and use of matrices.

The next chapter is concerned with determinants.

14 Determinants – functions revisited and a new notation

Objectives	This chapter provides a useful notation when working with certain kinds of arrays. Its aims are to

- define a determinant, and give rules for describing its expansion
- survey the properties of determinants
- provide a formalism for solving linear equations that arise in optimization problems in chemistry and physics
- demonstrate how matrix inverses are derived
- introduce a compact notation in preparation for describing some manipulations in vector algebra (developed in the next chapter)

14.1 The determinant of a square matrix

Particular sums of signed products of real numbers (or other objects) occur frequently in certain kinds of problem, and it has been found convenient to introduce a compact notation for such sums. The objects themselves, which are very often real numbers that can be identified with the elements of a square matrix, can be manipulated according to a precise prescription to yield a single number – the determinant of the matrix. Thus, for example, the 2×2 matrix

$$\begin{pmatrix} a & b \\ c & d \end{pmatrix}$$

has an associated determinant of order two, written as

$$\det \mathbf{A} = \begin{vmatrix} a & b \\ c & d \end{vmatrix} = (ad - bc),$$

which has the value $(ad - bc)$. Determinants should not be confused with matrices for, if a, b, c, d are real numbers, then

$$\begin{vmatrix} a & b \\ c & d \end{vmatrix}$$

is a real number. Thus we see that, in the situation considered here, where the elements of the matrix are numbers, the determinant can be defined in terms of an operation on the elements of a matrix to produce a value by means of a suitable prescription; that is, the determinant defines a function with a domain containing square matrices.

▷

Often called a determinant of *order* three.

The formalism may be extended to the evaluation of a 3×3 determinant of the form ▷

$$\begin{vmatrix} a & b & c \\ d & e & f \\ g & h & k \end{vmatrix},$$

by expanding it in terms of 2×2 determinants according to the rule

$$\begin{vmatrix} a & b & c \\ d & e & f \\ g & h & k \end{vmatrix} = a \begin{vmatrix} e & f \\ h & k \end{vmatrix} - b \begin{vmatrix} d & f \\ g & k \end{vmatrix} + c \begin{vmatrix} d & e \\ g & h \end{vmatrix}$$

$$= aek - ahf - bdk + bfg + cdh - ceg. \tag{14.1}$$

▷

For example, $c(dh) = (cd)h$, and $dch = dhc$, respectively.

Notice that the ordering of the symbols in each triple product is not significant, as multiplication is both associative and commutative ▷.

Here, the elements of the *first row* are taken in turn and multiplied by the determinant of what is left of the array when the row and column containing the element in question is crossed out; the products so obtained are then summed with the inclusion of a sign $(-1)^{row+col}$, where *row* (having the value 1) and *col* (taking the values 1, 2, 3) are indices designating the row and column number. For example, since b is in the first row and second column, the sign associated with this element is $(-1)^{1+2} = (-1)^3 = (-1)$.

The rule just described for expanding a determinant is applicable to any row or column. Thus, on expanding the third-order determinant above from the second column, we obtain

$$-b \begin{vmatrix} d & f \\ g & k \end{vmatrix} + e \begin{vmatrix} a & c \\ g & k \end{vmatrix} - h \begin{vmatrix} a & c \\ d & f \end{vmatrix} = -bdk + bfg + aek - ceg - ahf + cdh.$$

Problem 14.1

Evaluate

$$\begin{vmatrix} 1 & 3 & 0 \\ 2 & 6 & 4 \\ -1 & 0 & 2 \end{vmatrix}, \quad \begin{vmatrix} 1 & 2 & -1 \\ 3 & 6 & 0 \\ 0 & 4 & 2 \end{vmatrix}, \quad \begin{vmatrix} 1 & 3 & 0 \\ 1 & 3 & 2 \\ -1 & 0 & 2 \end{vmatrix}, \quad \begin{vmatrix} 1 & 1 & 0 \\ 1 & 1 & 1 \\ -1 & 0 & 1 \end{vmatrix}.$$

The discussion so far has been along the lines of a purely mechanical exposition in terms of rules for expanding a determinant of order two or three. It is clear that a determinant of arbitrary order n can be expanded in terms of determinants of order $n - 1$ which, in turn, can be treated in the same way recursively until a linear combination of 2×2 determinants is reached – all of which are readily evaluated. The only problem with this approach is that it becomes exceedingly tedious, because there are $n!/2$ such 2×2 determinants in the resulting expansion of a determinant of order n ▷. However, in this approach, a considerable amount of effort can be spared if the initial row or column selected is the one with the greatest number of zeros; also, the rule for determining the signs of the product terms in the expansion at any stage is easy to remember in the form

▷

The expansion of a 10×10 determinant, for example, yields 1 814 400 determinants of order two.

$$\begin{vmatrix} + & - & + & - & + & \cdots \\ - & + & - & + & \cdots & \cdots \\ + & - & + & \cdots & \cdots & \cdots \\ - & + & \cdots & \cdots & \cdots & \cdots \\ + & \cdots & \cdots & \cdots & \cdots & \cdots \end{vmatrix}$$

In practice, a given determinant can be transformed into a more manageable form, by using the following properties of determinants to introduce as many zero elements as possible before any expansion is attempted.

14.2   Properties of determinants

The properties are stated here without proof, because the main objective is to provide the means for working efficiently and effectively with determinants. All the properties of interest relate to manipulations of rows or columns:

▷

This implies that a determinant with two identical rows (columns) is zero.

(1) interchanging all rows and columns leaves the value of a determinant unaltered ($\det \mathbf{A} = \det \mathbf{A}^T$);
(2) interchanging two rows (columns) of a determinant changes its sign ▷;
(3) removing a constant factor c, from each element of any one row (column) yields c times a new determinant, with each element of the chosen row divided by c (this result is used in reverse to absorb a constant multiplying a given determinant by multiplying the elements of any chosen row (column) by that factor);
(4) a constant multiple of one row (column) added to (or subtracted from) another row (column) leaves the value of a determinant unaltered.

Problem 14.2

Show that

(a) $\begin{vmatrix} 4 & 2 & 2 \\ 2 & 4 & 2 \\ 2 & 2 & 4 \end{vmatrix} = 8 \begin{vmatrix} 2 & 1 & 1 \\ 1 & 2 & 1 \\ 1 & 1 & 2 \end{vmatrix} = 8 \begin{vmatrix} 0 & 0 & 1 \\ -1 & 1 & 1 \\ -3 & -1 & 2 \end{vmatrix} = 32,$

indicating which properties of determinants are used in each step.
(b) Rework Problem 14.1, using the properties of determinants to simplify each determinant before expansion.

$\triangleright$

Subtract row 2 from row 1 and then row 3 from row 2, and take out common factors from rows 1 and 2.

Problem 14.3

Use the row manipulation properties $\triangleright$ of determinants to show that

$$\begin{vmatrix} 1 & x & x^2 \\ 1 & y & y^2 \\ 1 & z & z^2 \end{vmatrix}$$

has the value $(x - y)(y - z)(z - x)$.

Worked example

14.1   Evaluate

$$\begin{vmatrix} \cos\theta & \sin\theta & 0 \\ -\sin\theta & \cos\theta & 0 \\ 0 & 0 & 1 \end{vmatrix}.$$

Solution   Expanding from the elements of the third row, using the appropriately modified form of Equation (14.1) and the sign convention above:

$$\begin{vmatrix} \cos\theta & \sin\theta & 0 \\ -\sin\theta & \cos\theta & 0 \\ 0 & 0 & 1 \end{vmatrix} = 0 \begin{vmatrix} \sin\theta & 0 \\ \cos\theta & 0 \end{vmatrix} - 0 \begin{vmatrix} \cos\theta & 0 \\ -\sin\theta & 0 \end{vmatrix}$$

$$+ 1 \begin{vmatrix} \cos\theta & \sin\theta \\ -\sin\theta & \cos\theta \end{vmatrix}$$

$$= \begin{vmatrix} \cos\theta & \sin\theta \\ -\sin\theta & \cos\theta \end{vmatrix} = \cos^2\theta + \sin^2\theta = 1.$$

$\triangleright$

From property 1.

Another useful strategy, involving row (column) manipulation, is to transform the given determinant of order n into a lower (or upper) $\triangleright$ triangular form

$$\begin{vmatrix} a_1 & 0 & 0 & \cdots & 0 \\ b_1 & b_2 & 0 & \cdots & 0 \\ c_1 & c_2 & c_3 & \cdots & 0 \\ \vdots & \vdots & \vdots & \ddots & \vdots \\ w_1 & w_2 & w_3 & \cdots & w_n \end{vmatrix}.$$

Its value is then given by the product $a_1 b_2 c_3 \cdots w_n$, formed from the elements on the diagonal – a result which is confirmed by expanding the determinant from the first row repeatedly.

Worked example

14.2   Evaluate

$$\begin{vmatrix} x & y & y \\ y & x & y \\ y & y & x \end{vmatrix}.$$

Solution

$$\begin{vmatrix} x & y & y \\ y & x & y \\ y & y & x \end{vmatrix} \xrightarrow{r_1 \to r_1 + r_2 + r_3} \begin{vmatrix} x+2y & x+2y & x+2y \\ y & x & y \\ y & y & x \end{vmatrix}$$

$$\xrightarrow{c_2 \to c_2 - c_1} \begin{vmatrix} x+2y & 0 & x+2y \\ y & x-y & y \\ y & 0 & x \end{vmatrix}$$

$$\xrightarrow{c_3 \to c_3 - c_1} \begin{vmatrix} x+2y & 0 & 0 \\ y & x-y & 0 \\ y & 0 & x-y \end{vmatrix} = (x+2y)(x-y)^2 .$$

Problem 14.4

Prove that

$$\begin{vmatrix} x & y & y & y & y \\ y & x & y & y & y \\ y & y & x & y & y \\ y & y & y & x & y \\ y & y & y & y & x \end{vmatrix} = (x-y)^4 (4y+x)$$

using the method given in Example 14.2 in order to bring the determinant into lower (or upper) triangular form.

14.3   Determinants with functions as elements

As we have seen already in Example 14.1, it is possible to have elements that are functions: thus, instead of yielding a single value, the determinant provides a way of defining a new function of the variables appearing in the elements. If, for example, a second-order determinant has as elements the functions $f(x), g(x), h(x)$ and $k(x)$ in the form,

$$\begin{vmatrix} f(x) & g(x) \\ h(x) & k(x) \end{vmatrix},$$

then expansion of the determinant according to the rule yields the function $f(x) k(x) - g(x) h(x)$ of the single variable x. It is clear, moreover, that since the differentiation of this function yields

$$\frac{d}{dx}\{f(x) k(x) - g(x) h(x)\} = f'(x) k(x) - g'(x) h(x) + f(x) k'(x) - g(x) h'(x)$$

it is possible to rearrange this result to obtain the rule for differentiating the original determinant:

$$\frac{d}{dx}\begin{vmatrix} f(x) & g(x) \\ h(x) & k(x) \end{vmatrix} = \begin{vmatrix} f'(x) & g'(x) \\ h(x) & k(x) \end{vmatrix} + \begin{vmatrix} f(x) & g(x) \\ h'(x) & k'(x) \end{vmatrix}.$$

Problem 14.5

Find the derivative of the determinant $\begin{vmatrix} 1 & \sin x \\ x & x^2 \end{vmatrix}$.

Problem 14.6

(a) Evaluate

$$\begin{vmatrix} \cos\theta & \sin\theta & 0 \\ \sin\theta & \cos\theta & 0 \\ 0 & 0 & 1 \end{vmatrix}.$$

(b) Find the first two derivatives of the determinant with respect to θ.
(c) Determine the values of θ for which the determinant has a maximum value.

The next three problems involve working with determinants which are functions of a single variable; but, because the determinant is required to have the value zero in both cases ▷, there are a finite number of values of the variable for which the equation is true.

▷
Such *secular determinants* occur widely in modelling bonding and vibrational motions in molecules, and are discussed more fully in Chapter 16.

Problem 14.7

For the matrix $\mathbf{C} = (3\ 1\ 2)$,

(a) construct the matrices $\mathbf{A} = \mathbf{CC}^T$, $\mathbf{B} = \mathbf{C}^T\mathbf{C}$
(b) find the values of λ such that $\det(\mathbf{A} - \lambda\mathbf{E}) = 0$ and $\det(\mathbf{B} - \lambda\mathbf{E}) = 0$, where $\mathbf{E}$ is the unit matrix of appropriate order.

▷
See also Problem 16.2.

Problem 14.8

The optimization of the form of the Hückel π-molecular orbitals for the cyclopropenyl cation leads to the determinantal equation ▷

$$\begin{vmatrix} \alpha - \epsilon & \beta & \beta \\ \beta & \alpha - \epsilon & \beta \\ \beta & \beta & \alpha - \epsilon \end{vmatrix} = 0,$$

for evaluating the molecular orbital energies, ϵ, in terms of α and β – the (fixed) parameters of the model.

(a) Expand the determinant and show that the permitted values of ϵ must satisfy the polynomial equation

$$\epsilon^3 - 3\alpha\epsilon^2 + 3\epsilon(\alpha - \beta)(\alpha + \beta) - (\alpha + 2\beta)(\alpha - \beta)^2 = 0.$$

(b) Given that the permitted values of ϵ may be written in the form $\epsilon = \alpha + \beta x$, substitute for ϵ in the polynomial equation, and show that x ▷ lies in the solution set of

$$x^3 - 3x - 2 = 0. \tag{14.2}$$

(c) Determine one element, λ_1, of the solution set of Equation (14.2) by inspection. Try $x = 0$, ± 1, ± 2.

(d) Given that $x = \lambda_1$ is a solution, then $(x - \lambda_1)$ is a factor of the cubic equation: that is,

$$(x - \lambda_1)(x^2 + bx + c) = 0,$$

where b, c are to be determined. Multiply out the left-hand side of the equation and, by comparing coefficients with the cubic equation in part (b), identify b and c.

(e) Solve the quadratic equation $x^2 + bx + c = 0$ to obtain the other two elements of the solution set of the original cubic equation ▷.

▷
Positive, zero, or negative values of x correspond to bonding, non-bonding, or antibonding molecular orbitals, respectively.

▷
Analogous calculations for buta-1,3-diene and cyclobutadiene are given in Atkins, Section 14.9.

Problem 14.9

Let

$$\mathbf{B} = \begin{pmatrix} 1 & 2 \\ 2 & 1 \end{pmatrix}; \quad \mathbf{G} = \begin{pmatrix} 1 \\ -1 \end{pmatrix}; \quad \mathbf{E} = \begin{pmatrix} 1 & 0 \\ 0 & 1 \end{pmatrix}.$$

(a) Show that $\mathbf{BG} = \lambda\mathbf{G}$, identifying the value of λ.
(b) Evaluate $\det(\mathbf{B} - \lambda\mathbf{E}) = 0$ to determine another value of λ.

(c) For this second value of λ, substitute

$$\mathbf{F} = \begin{pmatrix} p \\ q \end{pmatrix}$$

into $\mathbf{BF} = \lambda\mathbf{F}$ and determine p in terms of q. Use the condition $\mathbf{F}^T\mathbf{F} = 1$ to evaluate p (take the positive square root), and hence determine $\mathbf{F}$.

(d) Evaluate $\mathbf{F}^T\mathbf{G}$.

Determinantal functions of more than one variable are widespread in quantum chemistry, where they provide a very compact and convenient way of expressing wavefunctions for atoms or molecules containing two or more electrons. For such wavefunctions, each element of the determinant is a function of one spin and three space coordinates of an electron. The determinant, which yields a sum of signed products on expansion, is therefore a function of the space–spin coordinates of all the electrons. For example, in the case of the ground state of the helium atom, where the $1s$ atomic orbital is occupied by electrons with α- and β-spin, the determinantal wavefunction, $\Psi(1,2)$, is given by

$$\begin{vmatrix} 1s(1)\alpha(1) & 1s(1)\beta(1) \\ 1s(2)\alpha(2) & 1s(2)\beta(2) \end{vmatrix} = 1s(1)\alpha(1)1s(2)\beta(2) - 1s(1)\beta(1)1s(2)\alpha(2)$$

$$= 1s(1)1s(2)\{\alpha(1)\beta(2) - \alpha(2)\beta(1)\},$$

where $1s(1)$ signifies the $1s$ atomic orbital function for the electron labelled '1', and this label is a shorthand for the three spatial coordinates defining the position of the electron; $\alpha(1)$ and $\beta(1)$ are the spin functions for electron 1 associated with the 'spin-up' and 'spin-down' states of the electron.

The determinant of the spin orbitals in its expanded form is seen to be antisymmetric with respect to the exchange of electron labels 1 and 2 (a necessary property of electronic wavefunctions) $\triangleright$:

$\triangleright$
See Atkins, p. A32.

$$1s(1)1s(2)\{\alpha(1)\beta(2) - \alpha(2)\beta(1)\}(1 \leftrightarrow 2) \longrightarrow$$
$$1s(1)1s(2)\{\alpha(2)\beta(1) - \alpha(1)\beta(2)\} = -\Psi(1,2).$$

Determinants are also used to represent wavefunctions for systems containing more than two electrons but, in these cases, the properties of determinants are used to avoid explicit expansion of the wavefunction.

14.4   Notation

It is very convenient when dealing with determinants in general form to have a systematic way of referring to specific elements. The widely accepted notation is to use the row index notation introduced for labelling the elements of a matrix. Thus, the elements of a determinant are given a name together with row and column subscripts:

$$\det \mathbf{A} = \begin{vmatrix} a_{11} & a_{12} & \cdots & a_{1n} \\ a_{21} & a_{22} & \cdots & a_{2n} \\ a_{31} & a_{32} & \cdots & a_{3n} \\ \vdots & \vdots & \ddots & \vdots \\ a_{n1} & a_{n2} & \cdots & a_{nn} \end{vmatrix}$$

designates an $n \times n$ determinant.

The expansion of the 3×3 determinant (Equation (14.1)), for example, is now written in row–column notation as

$$\det \mathbf{A} = \begin{vmatrix} a_{11} & a_{12} & a_{13} \\ a_{21} & a_{22} & a_{23} \\ a_{31} & a_{32} & a_{33} \end{vmatrix}$$

$$= a_{11} \begin{vmatrix} a_{22} & a_{23} \\ a_{32} & a_{33} \end{vmatrix} - a_{12} \begin{vmatrix} a_{21} & a_{23} \\ a_{31} & a_{33} \end{vmatrix} + a_{13} \begin{vmatrix} a_{21} & a_{22} \\ a_{31} & a_{32} \end{vmatrix}. \qquad (14.3)$$

14.5   Cofactors of determinants

The cofactor A^{ij} of det $\mathbf{A}$ is obtained by deleting the ith row and jth column of the determinant to form an $(n-1) \times (n-1)$ determinant, which is then multiplied by $(-1)^{i+j}$.

Worked example

14.3   Write down A^{21} for the 3×3 determinant given in Equation (14.3).

Solution
$$A^{21} = (-1)^3 \begin{vmatrix} a_{12} & a_{13} \\ a_{32} & a_{33} \end{vmatrix} = - \begin{vmatrix} a_{12} & a_{13} \\ a_{32} & a_{33} \end{vmatrix}.$$

> **Problem 14.10**
>
> Determine the remaining eight cofactors associated with the determinant in Equation (14.3).

It is important to note that, although a determinant remains unchanged in value under the permitted row or column manipulations, the cofactor is not an invariant: for example, while subtracting column 1 from column 3 does not change the *value* of a determinant:

$$\begin{vmatrix} a & b & c \\ d & e & f \\ g & h & k \end{vmatrix} = \begin{vmatrix} a & b & c-a \\ d & e & f-d \\ g & h & k-g \end{vmatrix},$$

the respective values of A^{21} are changed by the column manipulation, from $-(bk - ch)$ to $-(bk - ch + ah - bg)$. Thus each determinant has its own characteristic set of cofactors.

Problem 14.11

(a) Evaluate A^{11}, A^{22}, A^{12}, A^{21} for the determinant $\begin{vmatrix} 1 & 3 & 0 \\ 2 & 6 & 4 \\ -1 & 0 & 2 \end{vmatrix}$.

(b) Evaluate A^{12} for the determinant in Problem 14.4.

Expanding a determinant in terms of cofactors

Using the results from Example 14.3 and Problem 14.10, the expansion from the first row of the 3×3 determinant given in Equation (14.3) takes the form

$$\det \mathbf{A} = \sum_{j=1}^{3} a_{1j} A^{1j}.$$

Similarly, expansion from the ith row or kth column yields the alternative expressions for $\det \mathbf{A}$ of

$$\sum_{j=1}^{3} a_{ij} A^{ij} \quad \text{and} \quad \sum_{j=1}^{3} a_{jk} A^{jk}, \tag{14.4}$$

respectively, where j is a counting index.

The cofactor expansion of an $n \times n$ determinant follows immediately:

$$\det \mathbf{A} = \sum_{j=1}^{n} a_{ij} A^{ij} = \sum_{j=1}^{n} a_{jk} A^{jk}.$$

If a determinant is expanded using either the elements from the ith row with the cofactors of the kth row, or the elements from the jth column with the cofactors of the kth column, then the result is zero in both cases:

$$\sum_{j=1}^{n} a_{ij} A^{kj} = 0 \text{ if } k \neq i, \tag{14.5}$$

$$\sum_{i=1}^{n} a_{ij} A^{ik} = 0 \text{ if } k \neq j. \tag{14.6}$$

These expressions are termed expansions in alien cofactors.

> **Problem 14.12**
>
> Use the cofactors in Example 14.3 and Problem 14.10 to demonstrate the truth of Equations (14.5), (14.6) for the choices $i = 1$, $k = 2$ (Equation (14.5)) and $j = 2$, $k = 3$ (Equation (14.6)).

14.6   Matrices revisited

This concluding section is concerned with the construction of the inverse of a square matrix, and its subsequent use in the solution of sets of simultaneous equations.

> **Problem 14.13**
>
> For the matrix
>
> $$\mathbf{A} = \begin{pmatrix} a_{11} & a_{12} \\ a_{21} & a_{22} \end{pmatrix},$$
>
> (a) show that the matrix of cofactors, $\mathbf{B}$, of $\det \mathbf{A}$ is given by
>
> $$\mathbf{B} = \begin{pmatrix} A^{11} & A^{12} \\ A^{21} & A^{22} \end{pmatrix} = \begin{pmatrix} a_{22} & -a_{21} \\ -a_{12} & a_{11} \end{pmatrix};$$
>
> (b) compute $\mathbf{B}^T \mathbf{A}$ and show, using Equations (14.4), (14.6), that the product matrix takes the form
>
> $$\mathbf{B}^T \mathbf{A} = \det \mathbf{A} \begin{pmatrix} 1 & 0 \\ 0 & 1 \end{pmatrix} \Rightarrow \frac{1}{\det \mathbf{A}} \mathbf{B}^T \mathbf{A} = \mathbf{E}_2,$$
>
> where $\mathbf{E}_2$ is the unit matrix of order 2;
>
> (c) demonstrate that $\mathbf{A}\mathbf{B}^T$ also yields $\mathbf{E}_2$.

The inverse matrix

The last results in Problem 14.13 are important, because it shows that there is a matrix

$$\frac{1}{\det \mathbf{A}} \mathbf{B}^T \tag{14.7}$$

which, when either postmultiplied or premultiplied by $\mathbf{A}$, yields the identity matrix. Equation (14.7) defines the *inverse* of matrix $\mathbf{A}$, and is usually designated as $\mathbf{A}^{-1}$. Clearly, the inverse matrix exists only for a square matrix and, also, if $\det \mathbf{A} \neq 0$.

Problem 14.14

Use the cofactors evaluated for the determinant in Problem 14.11(a) to construct the transpose of the cofactor matrix, and hence determine the inverse of the matrix

$$\mathbf{A} = \begin{pmatrix} 1 & 3 & 0 \\ 2 & 6 & 4 \\ -1 & 0 & 2 \end{pmatrix}.$$

Solution of simultaneous equations

▷
Inhomogeneous means that the number on the right-hand side of each equation is non-zero.

We now make a short excursion into the problem of solving sets of simultaneous equations which are described as *linear inhomogeneous* ▷ equations. Equations of this form arise naturally in group theory, for example, where we are interested in how the coordinates of a point are transformed under a symmetry operation. In the special case, when the right-hand side of each equation is zero, we have a set of *homogeneous* equations, and such equations play a central role in the modelling of the electronic structure of atoms and molecules and also in the related problem (for molecules) of determining the frequencies of the permitted nuclear motions.

The solution of three linear inhomogeneous equations in the three unknowns x, y, z is developed in the next problem with the aid of the matrix notation.

Problem 14.15

Write each of the following sets of linear inhomogeneous equations

(i) $\quad 2x - y + z = 2$ (ii) $\quad -3x + 6y - 11z = -1$
$\quad\quad x + 2y - z = 3$ $\quad\quad\quad 3x - 4y + 6z = 2$
$\quad\quad 3x + y + 2z = -1$ $\quad\quad\quad 4x - 8y + 13z = 3$

(iii) $\quad x \cos\theta - y \sin\theta = x'$
$\quad\quad x \sin\theta + y \cos\theta = y'$
$\quad\quad\quad\quad\quad\quad z = z'$

in the matrix form $\mathbf{Ax} = \mathbf{b}$, where $\mathbf{A}$ is the matrix of coefficients;

$$\mathbf{x} = \begin{pmatrix} x \\ y \\ z \end{pmatrix},$$

and $\mathbf{b}$ is the column matrix of coefficients appearing on the right-hand side of the given set of equations. For each set of equations, with given $\mathbf{A}$ and $\mathbf{b}$, determine

$\triangleright$
If $\mathbf{Ax} = \mathbf{b}$ then
$\mathbf{A}^{-1}\mathbf{Ax} = \mathbf{A}^{-1}\mathbf{b} \Rightarrow$
$\mathbf{Ex} = \mathbf{A}^{-1}\mathbf{b} \Rightarrow \mathbf{x} = \mathbf{A}^{-1}\mathbf{b}.$

(a) $\mathbf{A}^{-1}$, and verify that $\mathbf{A}^{-1}\mathbf{A} = \mathbf{E}_3$;
(b) $\mathbf{x}$, in the form $\mathbf{A}^{-1}\mathbf{b}$ $\triangleright$;
(c) check that the values obtained for each set of x, y and z values do, in fact, satisfy the respective original equations.

Problem 14.15(c) is an especially important example in the use of matrices because $\mathbf{A}$ is the matrix representation of a physical rotation about the z-axis. This rotation describes how the new coordinates of a point in space are related to the coordinates of a point prior to the rotation taking place – an important aspect of symmetry theory, an outline of which has been given towards the end of Chapter 13. The use of the matrix inverse allows us, therefore, to compute coordinates of a transformed point prior to a symmetry transformation (here a rotation), since $\mathbf{x}' = \mathbf{Ax} \Rightarrow \mathbf{x} = \mathbf{A}^{-1}\mathbf{x}'$.

$\triangleright$
A solution other than
$x = y = z = \cdots = 0.$

If the matrix $\mathbf{b}$ is null, then the set of (homogeneous) equations admits of a non-trivial solution $\triangleright$ only if det $\mathbf{A} = 0$. If the determinant of the coefficients is non-zero, then the equations are inconsistent and do not admit of a solution.

Problem 14.16

Find the values of k for which the equations

$$kx - 4y + z = 0, \quad 2x - 2y - kz = 0, \quad x + 2y - 2z = 0$$

possess a non-trivial solution.
Solve the equations for each value of k, given that $x^2 + y^2 + z^2 = 1$.

Summary: The aims set out at the beginning of this chapter are now achieved; applications of the notation, and the use of the properties of determinants, will form integral parts of later chapters – especially in Chapter 16 on the eigenvalue problem. In the meantime, we now proceed to review the use and properties of vectors.

15 Vectors – a formalism for directional properties

Objectives	This chapter

This chapter

- develops a notation for handling properties that are characterized by a magnitude and a direction
- provides some basic insights into three-dimensional geometry
- links the geometrical view of vectors with the formalism of matrices
- reviews chemical applications – especially in the study of the solid state

For the most part (certainly in the introductory phases of the subject), chemistry is concerned with properties that are characterized by a single number (a scalar) with appropriate units attached: for example, amount of substance, concentration, acidity, enthalpy, entropy, rate of a reaction, electronegativity, wavelength, etc. There are areas of chemistry, however, where direction plays a central role in defining the property or associated characteristic. Typical examples are the response of a molecular or solid state system to applied electric or magnetic fields; the direction of a dipole moment; the vibrational motions of atoms in a molecular species; the shapes of crystalline unit cells; transition states and the associated minimum energy paths involving reacting species; diffusion and other transport processes in liquids and solids (electrical conductivity), and the treatment of surface phenomena (catalysis).

In all of these situations we are concerned with a geometrical problem in some form or other as direction plays a key role. Since vectors are defined in terms of a *magnitude* and a *direction*, they provide a natural structure for describing directional chemical properties.

In motivating the ideas behind vectors, we take the easy way of appealing to intuition by developing an approach in terms of vectors in three-dimensional space. We shall then be in the position to extend the principal results to more than three dimensions because, in quantum chemistry and spectroscopy, we are concerned with the motions of electrons and nuclei, each of which requires three coordinates to define its position. The nuclear motions in ammonia, for example, are determined by specifying twelve coordinates; these in turn define a vector in a twelve-dimensional space. Within this space there is a subspace of six dimensions, trajectories in which the possible internal vibrational motions of the molecule are defined. For the moment, however, we return to the more mundane situations in which we explore the use of vectors in two and three dimensions.

A vector in two- or three-dimensional space may be represented by means of a directed line segment, $\overrightarrow{OA}$ say, where O and A are the initial and final points of the line segment, the length of which represents the magnitude and an arrow the direction of the vector; a *unit* vector has unit magnitude. It is clear, therefore, that unless we are dealing with, for example, a force acting through a particular point, there is nothing special about where the directed line segment is placed in space, and any directed line segment of the same length (magnitude) which is parallel to the given vector represents the same vector; it is only a matter of convenience which line segment is selected to represent the vector.

15.1   Conventions

▷
In the form of a sans-serif letter

(1) Vectors are represented either by specifying the initial and final points of the directed line segment (as described above) or by giving them a name ▷. Thus, for example, $\overrightarrow{OA} \equiv$ a, $\overrightarrow{GH} \equiv$ b, ... The vector corresponding to $\overrightarrow{AO}$ is represented by $-$a , since only the direction is reversed; on the other hand, if the directed line segment $\overrightarrow{OA}$ is doubled in length, then the new vector is represented by 2 a , as shown in Figure 15.1.

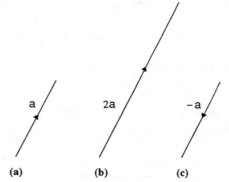

Figure 15.1
Line segment representations of the vectors: (a) a, (b) 2a, and (c) $-$a

(2) A unit vector in the direction of a is represented by

$$\hat{a} = \frac{a}{|a|}.$$ (15.1)

It therefore follows that $a = |a|\,\hat{a}$, where the magnitude $|a|$ (or just a) is the component of a in the direction of the unit vector $\hat{a}$.

15.2   Addition of vectors

The addition of the two vectors $\overrightarrow{OP} \equiv a$, $\overrightarrow{ST} \equiv b$, (Figure 15.2(a)) is determined by first translating one vector relative to the other so that their initial points coincide (Figure 15.2(b)); each vector is then replicated to form the parallelogram as shown in Figure 15.2(c). Thus, in terms of directed line segments, we have

$$\overrightarrow{OQ} = \overrightarrow{OP} + \overrightarrow{OR} \text{ (by definition)}$$
$$= \overrightarrow{OP} + \overrightarrow{PQ}$$

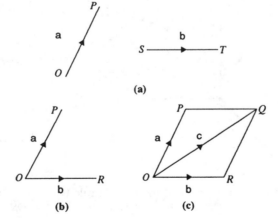

Figure 15.2
Three stages in the construction of the parallelogram for the sum of two vectors a and b

which may be written in terms of vector names as $c = a + b$, a result which is often interpreted in terms of the *triangle rule* (Figure 15.3).

Since from Figure 15.2(c) $\overrightarrow{RQ} + \overrightarrow{QP} = \overrightarrow{RP}$ ($\overrightarrow{QP}$ is in the opposite direction to $\overrightarrow{PQ}$), it is clear from Figure 15.2(c) that the directed line segment $\overrightarrow{RP}$, forming the other diagonal of the parallelogram, is the vector $a - b$.

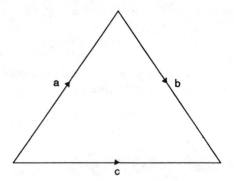

Figure 15.3
The triangle rule: c = a + b

15.3   Base vectors

The addition (and subtraction) of vectors has been described above in terms of a geometrical construction. This is quite tedious in practice, and it's much better to refer vectors to a common set of unit vectors (using repeated applications of the triangle rule): a procedure which reduces manipulations of vectors to algebraic operations. For example, in three-dimensional space, the three unit vectors $\hat{i}$, $\hat{j}$, $\hat{k}$ are conveniently taken to be directed along the positive (right-handed) Cartesian x, y and z axes, as indicated in Figure 15.4.

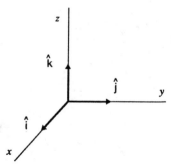

Figure 15.4
Unit factors $\hat{i}$, $\hat{j}$, $\hat{k}$ associated with a Cartesian coordinate system

▷
Reminder: $\overrightarrow{OR}$ is written in terms of a magnitude times a unit vector in the direction of the x-axis.

If $\overrightarrow{OP} \equiv r$, then using the triangle rule twice
$r = \overrightarrow{OQ} + \overrightarrow{QP} = (\overrightarrow{OR} + \overrightarrow{RQ}) + \overrightarrow{QP} = \overrightarrow{OR} + \overrightarrow{OS} + \overrightarrow{QP}$. But from Equation (15.1), as $\overrightarrow{OR} = |\overrightarrow{OR}|\hat{i} = x\hat{i}$, ▷ etc., it follows that $r = x\hat{i} + y\hat{j} + z\hat{k}$, where (x, y, z) are the coordinates of P or the *components* of the vector r referred to the base vectors $\hat{i}$, $\hat{j}$, $\hat{k}$. Furthermore, by the Pythagoras theorem for right-angled triangles,

$$|r|^2 \equiv r^2 = |\overrightarrow{OP}|^2 = |\overrightarrow{OQ}|^2 + |\overrightarrow{QP}|^2 = |\overrightarrow{OR}|^2 + |\overrightarrow{RQ}|^2 + |\overrightarrow{QP}|^2$$
$$= |\overrightarrow{OR}|^2 + |\overrightarrow{OS}|^2 + |\overrightarrow{QP}|^2,$$

and it follows that $|r| = r = (x^2 + y^2 + z^2)^{\frac{1}{2}}$.

Worked example

15.1 If $r = 2\hat{\imath} + 3\hat{\jmath}$, $s = \hat{\imath} - \hat{\jmath} + \hat{k}$, find (a) $r + s$, (b) $r - s$, (c) $|r + s|$.

Solution

$$r + s = (2\hat{\imath} + 3\hat{\jmath}) + (\hat{\imath} - \hat{\jmath} + \hat{k}) = 3\hat{\imath} + 2\hat{\jmath} + \hat{k}$$
$$r - s = (2\hat{\imath} + 3\hat{\jmath}) - (\hat{\imath} - \hat{\jmath} + \hat{k}) = \hat{\imath} + 4\hat{\jmath} - \hat{k}$$
$$|r + s| = \sqrt{3^2 + 2^2 + 1^2} = \sqrt{14}.$$

Problem 15.1

If $a = 2\hat{\imath} + \hat{\jmath} - \hat{k}$, $b = -\hat{\imath} + 4\hat{k}$, $c = \hat{\imath} + 2\hat{\jmath} + 3\hat{k}$, $d = 3\hat{\imath} - \hat{k}$, find

(a) $a + b$, (b) $a - b$, (c) $2a + 3b$, (d) $|a + b + c + d|$,
(e) $a/|a|$, (f) $|a| + |b|$, (g) $b + 0.5d$.

15.4   Vector multiplication

The *scalar* and *vector* products involve different binary combinations of vectors that are conveniently thought of as means of multiplication; just as with matrix algebra, though, division is not a meaningful binary operation. The scalar product is so called because the defining binary operation yields a scalar quantity (a magnitude, devoid of direction); the vector product, not surprisingly, yields a vector under its defining binary operation.

The scalar product

Consider two vectors a, b, with initial points made coincident, and where θ is the angle between a and b (Figure 15.5). The scalar product is defined as

$$a \cdot b = |a||b| \cos\theta = b \cdot a. \tag{15.2}$$

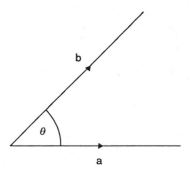

Figure 15.5
The angle between two vectors lying in a plane

▷

Here order of
multiplication is not
important.

It is clear that this binary operation is commutative, because the result of carrying out the operation defined by the scalar product is just a product of scalar quantities ▷. Furthermore, if b is replaced by a in Equation (15.2), then it follows that $a \cdot a = |a|^2$, and this operation provides a direct route for evaluating the magnitude of a vector. It is also worth noting from the trigonometrical properties of right-angled triangles that since $|a| \cos \theta$ is the component of a in the direction of b , and $|a| \cos (\pi/2 - \theta)$ is the component of a in the direction perpendicular to b , a given vector can always be replaced by a sum of two vectors at right angles to each other: a process that is termed *resolution*.

▷

Sometimes systems of
unit vectors appropriate
to other coordinate
systems are used.

Scalar products are easy to evaluate if the two vectors a, b are expressed in terms of the set of Cartesian base vectors $\hat{i}, \hat{j}, \hat{k}$ in the form ▷ $a = x_1 \hat{i} + y_1 \hat{j} + z_1 \hat{k}$, $b = x_2 \hat{i} + y_2 \hat{j} + z_2 \hat{k}$. The advantage of this approach is that only scalar products between the base vectors need to be evaluated, and this can be done once and for all. Furthermore, as the angles between pairs of different base vectors are all $90° \equiv \pi^c/2$, it follows from Equation (15.2) that

$$\hat{i} \cdot \hat{j} = \hat{i} \cdot \hat{k} = \hat{k} \cdot \hat{j} = 0 \,,$$

since $\cos (\pi/2) = 0$. The remaining scalar products between a base vector and itself involve substituting $\theta = 0$ in the defining formula

$$\hat{i} \cdot \hat{i} = \hat{j} \cdot \hat{j} = \hat{k} \cdot \hat{k} = 1 \,,$$

yielding $a \cdot b = x_1 x_2 + y_1 y_2 + z_1 z_2$.

If $a \cdot b = 0$ then a and b are said to be *orthogonal*, which is a better description than 'at right angles', because, when working in more than three dimensions, the conventional geometrical concepts of line and angle must be generalized.

Worked example

15.2   For the r, s given in Example 15.1,

(a) determine r · s, and the acute angle between r and s
(b) resolve r into a sum of two orthogonal vectors, one of which is in the direction of s .

Solution   (a) Since $r = 2\hat{i} + 3\hat{j}$, $s = \hat{i} - \hat{j} + \hat{k}$,

$$r \cdot s = (2\hat{i} + 3\hat{j}) \cdot (\hat{i} - \hat{j} + \hat{k}) = 2(\hat{i} \cdot \hat{i}) - 3(\hat{j} \cdot \hat{j}) = -1 \,.$$

But $|r| = \sqrt{2^2 + 3^2} = \sqrt{13}$, $|s| = \sqrt{1^2 + 1^2 + 1^2} = \sqrt{3}$ and, using the definition of scalar product in Equation (15.2), it follows that

$$r \cdot s = -1 = \sqrt{13} \cdot \sqrt{3} \cos \theta \Rightarrow \theta = \cos^{-1} \left(\frac{-1}{\sqrt{39}} \right) = 91.5° \,.$$

The angle of smallest magnitude between the two vectors (the acute angle) is therefore 88.5°.

(b) The component of r in the direction of s is $|r| \cos \theta = -1/\sqrt{3}$ and, since $\hat{s} = (\hat{i} - \hat{j} + \hat{k})/\sqrt{3}$, the vector, w, in the direction of s is $-(\hat{i} - \hat{j} + \hat{k})/3$. Hence, applying the triangle rule r = w + v, where v is the vector perpendicular to w, we find

$$v = r - w = \frac{7}{3}\hat{i} + \frac{8}{3}\hat{j} + \frac{1}{3}\hat{k}.$$

Problem 15.2

If $a = \hat{i} + 5\hat{k}$, $b = 3\hat{i} + \hat{j} + 4\hat{k}$, $c = \hat{i} - \hat{j} + 2\hat{k}$, find

(a) $a \cdot b$,   (b) $a \cdot (b + c)$,   (c) $a \cdot (b - c)$,   (d) $2b \cdot (3a - 4c)$
(e) the angle between $(b - c)$ and a
(f) the value of λ such that b is orthogonal to $v = 4\hat{i} + \lambda\hat{j}$.

Problem 15.3

If the origin of coordinates is taken on the carbon atom in the methane molecule, the vectors r_i, $(i = 1, 2, 3, 4)$, having components $(1, 1, 1)$, $(-1, -1, 1)$, $(-1, 1, -1)$, $(1, -1, -1)$, are each directed towards one hydrogen atom, H_i. If the C–H_i bond distance is represented by R,

(a) determine the unit vectors $\hat{r}_i$, and hence write down the four C–H_i bond vectors, h_i, each having magnitude R;
(b) use Equation (15.2) to determine the angle between h_1 and h_2;
(c) determine $h_1 - h_2$, and hence express the distance between H_1 and H_2, $|h_1 - h_2|$, in terms of R.

Problem 15.4

The Cartesian coordinates in Å ($1 \text{ Å} = 10^{-10}$ m) of the nuclei in chloroethane are held in a molecular geometry database ▷ in the form

C	1	−0.002777	−0.332535	−0.659256
C	2	−0.010651	−0.318436	0.863647
Cl	3	−0.003036	1.367386	−1.235672
H	4	0.942474	−0.796371	−1.019989
H	5	−0.942520	−0.799911	−1.029709
H	6	−0.010712	−1.360733	1.253998
H	7	−0.918579	0.207565	1.234772
H	8	0.891449	0.210907	1.244095

where each row consists of an atom name, a label index, and the x, y, z coordinates of the nucleus.

▷
A repository for data and other information held on a computer.

▷
All atoms are bonded
either to C_1 or to C_2.

Calculate

(a) the distances between bonded atoms in pm ▷
(b) the $C_2\text{–}\widehat{C_1}\text{–}Cl$ and $H_8\text{–}\widehat{C_2}\text{–}H_7$ bond angles.

The vector product

As already indicated above, the vector product between two vectors yields a vector, the prescription for which is given by

$$a \times b = |a||b| \sin\theta\,\hat{n} = -b \times a.$$ (15.3)

The resulting vector is of magnitude $|a||b|\sin\theta$, and lies in the direction of the unit vector $\hat{n}$, which is perpendicular to the plane containing a and b. The sense of $\hat{n}$ can be in one of two directions, and the particular direction implied by the prescription is determined by the requirement that a, b and $\hat{n}$, in order, form a right-handed system, as indicated in Figure 15.6.

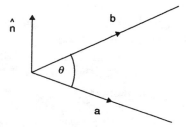

Figure 15.6
The direction of the unit vector $\hat{n}$ for the vector product $a \times b$

Equation (15.3) shows two other important features: first, this binary operation on vectors is not commutative and, second, substituting a = b in the defining relation, yields $a \times a = -a \times a \Rightarrow a \times a$ is the null vector ▷.

▷
A vector of zero magnitude.

▷
Reminder: $\hat{i},\hat{j},\hat{k}$ are right-hand related.

Problem 15.5

Use the rule for determining the vector product given in Equation (15.3) to demonstrate that ▷

(a) $\hat{i} \times \hat{i} = \hat{j} \times \hat{j} = \hat{k} \times \hat{k} = 0$, and
(b) $\hat{i} \times \hat{j} = \hat{k},\ \hat{i} \times \hat{k} = -\hat{j},\ \hat{k} \times \hat{j} = -\hat{i}$

Problem 15.6

Use the definitions of a, b and c in Problem 15.2 to determine

(a) $a \times b$, (b) $b \times a$, (c) $|b \times c|$, (d) $(a - b) \times (c - a)$, (e) $a \cdot (b \times c)$

▷
Or a trapezium in
general.

Problem 15.7

The area of the parallelogram ▷ $OPQR$ (Figure 15.7), with a and b as adjacent sides, is given by $\frac{1}{2}\{|\overrightarrow{OR}| + |\overrightarrow{PQ}|\}h = \frac{1}{2}\{|a| + |b|\}h = |a|h$, where h is the perpendicular distance between $\overrightarrow{OR}$ and $\overrightarrow{PQ}$. Show that

(a) $h = |b|\sin\theta$, (b) the area of the parallelogram may be written as $|a \times b|$.

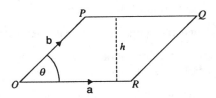

Figure 15.7
The parallelogram formed
from the vectors a and b

Problem 15.8

If $a = a_1\hat{i} + a_2\hat{j} + a_3\hat{k}$, $b = b_1\hat{i} + b_2\hat{j} + b_3\hat{k}$, $c = c_1\hat{i} + c_2\hat{j} + c_3\hat{k}$, show that

(a) $a \cdot b = a_1 b_1 + a_2 b_2 + a_3 b_3 = a^T b$, where **a** and **b** are column matrices with elements a_i and b_i ($i = 1, 2, 3$), respectively.

(b) $b \times c = \begin{vmatrix} \hat{i} & \hat{j} & \hat{k} \\ b_1 & b_2 & b_3 \\ c_1 & c_2 & c_3 \end{vmatrix}$ and $D = a \cdot (b \times c) = \begin{vmatrix} a_1 & a_2 & a_3 \\ b_1 & b_2 & b_3 \\ c_1 & c_2 & c_3 \end{vmatrix}$.

Hint: In finding D, expand the determinantal form of $b \times c$ from the first row, prior to taking the scalar product with a.

(c) $D = b \cdot c \times a = c \cdot a \times b$ from the properties of determinants.

(d) $a \times (b \times c) = \begin{vmatrix} \hat{i} & \hat{j} & \hat{k} \\ a_1 & a_2 & a_3 \\ D^{11} & D^{12} & D^{13} \end{vmatrix}$;

$(a \times b) \times c = -c \times (a \times b) = - \begin{vmatrix} \hat{i} & \hat{j} & \hat{k} \\ c_1 & c_2 & c_3 \\ D^{31} & D^{32} & D^{33} \end{vmatrix} \neq a \times (b \times c)$

where the D^{ij} are the cofactors associated with the first and third rows of D in (b) above.

There are important facets of different parts of the last problem that deserve some comment:

- In part (a), it is seen how the scalar product a·b of the *vectors* a and b can be written in terms of a product of a transposed column *matrix* of the components of a with the (column) *matrix* of the components of b – thus exhibiting another example of an isomorphism between vectors and their representation in terms of (column) matrices. Vector manipulations can therefore be replaced by matrix operations if the vectors are first mapped to matrices; then the appropriate matrix operation is carried out and, finally, the resulting matrix is mapped back to a vector. The scalar product a·b therefore maps to the matrix product $\mathbf{a}^T\mathbf{b}$.

- Determinants arise quite naturally with elements other than simple scalars.

- The determinantal form of the vector product provides the simplest route to the evaluation of a ×b .

- The vector product does not show the property of associativity as, clearly, $(a \times b) \times c \neq a \times (b \times c)$.

The vector product in chemistry

The vector product, in one form or other, arises in several different contexts: in atomic structure, in crystallography, and in the quantum mechanics of the interaction of electromagnetic radiation with matter. For example, in discussing the electronic motion in atoms and linear molecules, the *angular momentum* is an important physical property which is concerned with the moment of the linear momentum of the electron, with position vector r, about an axis of rotation. If the components of r are (x, y, z) and the components of linear momentum, p, are (mv_x, mv_y, mv_z), then the angular momentum is defined as the vector r × p, which has a direction perpendicular to the plane containing r and p ▷.

▷
For a particle circulating in an orbit lying in the *xy* plane, the direction of the angular momentum is perpendicular to the plane of the orbit.

The result proved in Problem 15.8(b) can be used to specify the volume, V, of a parallelepiped with adjacent sides a, b and c as $V = |a \cdot (b \times c)|$, a useful result in crystallography, since the basic unit cells of crystalline materials are parallelepipeds of different characteristic kinds.

| 15.5 | **Geometry in two and three dimensions** |

Straight lines, curves, planes and surfaces form an important constituent in the modelling of chemical phenomena. Very often the handling of these entities is carried out algebraically; however, with the power of the vector notation, it is a worthwhile exercise to see how some of these geometrical constructs are amenable to an alternative development.

The straight line

The ubiquitous straight line function is commonly perceived in two dimensions when we ascribe the algebraic form $y = mx + c$. Proceeding a little further in this direction, we know that a straight line is fully specified once two points, $(x_1, y_1), (x_2, y_2)$ on the line are given. Thus, the two equations $y_1 = mx_1 + c$ and $y_2 = mx_2 + c$ can be solved for m and c to obtain the formula in the alternative form

$$y = \left(\frac{y_2 - y_1}{x_2 - x_1}\right) x + \frac{y_1 x_2 - x_1 y_2}{x_2 - x_1} \equiv y = mx + c.$$

The use of vectors provides a more satisfying, and generalizable, formulation of the straight line function. Consider two vectors r_1, r_2 in the xy plane, with initial points at the origin and final points $(x_1, y_1), (x_2, y_2)$ lying on the required straight line. Using the triangle rule, it is seen that the vector $r_2 - r_1$ lies in the straight line. For any arbitrary point (x, y), the vector from the origin to the point is referred to as r. Thus, $(r - r_1) \times (r_2 - r_1)$ is a vector perpendicular to the plane of the form $\lambda \hat{k}$, that is,

$$\hat{k} \cdot (r - r_1) \times (r_2 - r_1) = \lambda.$$

Now if (x, y) lies on the line, so that $(r_2 - r_1)$ is parallel to $(r - r_1)$, then the vector product yields a null vector, and λ is zero. The use of the expression for D in Problem 15.8(b) thus enables the equation of the straight line to be written as

$$\hat{k} \cdot ((r - r_1) \times (r_2 - r_1)) = \begin{vmatrix} 0 & 0 & 1 \\ x - x_1 & y - y_1 & 0 \\ x_2 - x_1 & y_2 - y_1 & 0 \end{vmatrix}$$

$$= \begin{vmatrix} x - x_1 & y - y_1 \\ x_2 - x_1 & y_2 - y_1 \end{vmatrix} = 0.$$

Problem 15.9

(a) Show that the equation of a straight line in the xy plane may be written in the alternative form

$$\frac{x - x_1}{x_2 - x_1} = \frac{y - y_1}{y_2 - y_1}.$$

(b) Write down the equation of the straight line passing through the points $(2, -1)$ and $(3, 2)$, giving its slope and intercept on the y-axis.

In extending these ideas to three dimensions, there are no real differences in principle since the vectors each have appropriate x, y, z components. Thus, as before $(r - r_1) \times (r_2 - r_1) = 0$ and, on equating components on each side of this equation using the first result in Problem 15.8(b), the equations

$$\frac{x - x_1}{x_2 - x_1} = \frac{y - y_1}{y_2 - y_1} = \frac{z - z_1}{z_2 - z_1}$$

are obtained – a simple generalization of the result for two dimensions. Of more interest, however, is the way we define the equation of a plane, since this has no counterpart in two dimensions.

The plane

$\triangleright$

All lie in the plane.
The scalar product of two perpendicular vectors is zero.

For three vectors r_1, r_2, r_3, with initial points at the origin, the end-points lie in a plane. Given an arbitrary vector, r, with components (x, y, z), and end-point lying in the plane, then the three vectors $a = r - r_1$, $b = r - r_2$, $c = r - r_3$ are coplanar $\triangleright$ and $a \cdot (b \times c) = 0$ since $b \times c$ is perpendicular to the plane *and* to a. Thus, the equation of a plane is

$$\begin{vmatrix} a_1 & a_2 & a_3 \\ b_1 & b_2 & b_3 \\ c_1 & c_2 & c_3 \end{vmatrix} = \begin{vmatrix} x - x_1 & y - y_1 & z - z_1 \\ x - x_2 & y - y_2 & z - z_2 \\ x - x_3 & y - y_3 & z - z_3 \end{vmatrix} = 0, \tag{15.4}$$

where the components of r_i are (x_i, y_i, z_i).

Determinants revisited

The second determinant in Equation (15.4) is tedious to expand using the usual rules. In this sort of situation, where the elements are differences of two variables, it is more advantageous to expand from the first row as usual:

$$(x - x_1)\begin{vmatrix} y - y_2 & z - z_2 \\ y - y_3 & z - z_3 \end{vmatrix} - (y - y_1)\begin{vmatrix} x - x_2 & z - z_2 \\ x - x_3 & z - z_3 \end{vmatrix}$$
$$+ (z - z_1)\begin{vmatrix} x - x_2 & y - y_2 \\ x - x_3 & y - y_3 \end{vmatrix} = 0, \tag{15.5}$$

and then to reassemble the positive and negative terms in the form of two 3×3 determinants

$$\begin{vmatrix} x & y & z \\ x - x_2 & y - y_2 & z - z_2 \\ x - x_3 & y - y_3 & z - z_3 \end{vmatrix} - \begin{vmatrix} x_1 & y_1 & z_1 \\ x - x_2 & y - y_2 & z - z_2 \\ x - x_3 & y - y_3 & z - z_3 \end{vmatrix} = 0. \tag{15.6}$$

▷

If any determinant produced in this manner has two identical rows then, from the properties of determinants, its value is zero.

Problem 15.10

(a) Using the method of determinantal expansion given in Equations (15.5), (15.6), continue the expansion of Equation (15.4) from the second, and then from the third rows ▷, to show that the equation for a plane can be written in the form $Ax + By + Cz = D$, where

$$A = \begin{vmatrix} 1 & y_1 & z_1 \\ 1 & y_2 & z_2 \\ 1 & y_3 & z_3 \end{vmatrix}, \quad B = \begin{vmatrix} x_1 & 1 & z_1 \\ x_2 & 1 & z_2 \\ x_3 & 1 & z_3 \end{vmatrix}, \quad C = \begin{vmatrix} x_1 & y_1 & 1 \\ x_2 & y_2 & 1 \\ x_3 & y_3 & 1 \end{vmatrix},$$

$$D = \begin{vmatrix} x_1 & y_1 & z_1 \\ x_2 & y_2 & z_2 \\ x_3 & y_3 & z_3 \end{vmatrix}.$$

(b) As (x, y, z) are the components of the vector r, whose end-point lies in the plane, it is clear that since the end-points of r_i also lie in the plane then $Ax_i + By_i + Cz_i = D$. Show that if the vector n is defined according to $n = A\hat{i} + B\hat{j} + C\hat{k}$, then the equation of the plane may be written as

$$(r - r_i) \cdot n = 0.$$

It should be noted that the vector $(r - r_i)$ lies in the plane.

▷

Perpendicular in two or three dimensions.
▷
Alberty and Silbey, Section 23.3.

Problem 15.10 yields an important result, because it demonstrates that a plane is defined by two vectors: one in the plane, and the other, n, *normal* ▷ to the plane. Planes are important in X-ray diffraction studies of crystals, because their orientations, characterized by the intercepts on the x, y and z axes, are required for the full elucidation of the crystal structure ▷.

▷

For a more general discussion of crystal lattice types, see Alberty and Silbey, Section 23.2.

Problem 15.11

A crystal lattice generated by the base vectors
$a_1 = 3\hat{i}, a_2 = \hat{i} + 2\hat{j}, a_3 = \hat{i} + \hat{j} + \hat{k}$ enables a general lattice point to be given by $a = n_1 a_1 + n_2 a_2 + n_3 a_3$, where n_1, n_2, n_3 are integers ▷.

(a) Calculate the angles bewteen a_1 and a_2, a_1 and a_3, and a_2 and a_3.
(b) Use the results in Problem 15.8(b) to determine the volume, V, of the basic unit cell.
(c) If three new base vectors b_1, b_2, b_3, defined in terms of the unit vectors $\hat{i}, \hat{j}, \hat{k}$ according to $b_1 = x_1\hat{i} + y_1\hat{j} + z_1\hat{k}$, $b_2 = x_2\hat{i} + y_2\hat{j} + z_2\hat{k}$, $b_3 = x_3\hat{i} + y_3\hat{j} + z_3\hat{k}$, are constrained such that

$$b_i \cdot a_i = 1, b_i \cdot a_j = 0 (i \neq j)$$

write down and solve the equations for the set of (x_i, y_i, z_i) corresponding to each b_i.

(d) Explain why b_1 is of the form $k(a_2 \times a_3)$, and show by evaluating $a_1 \cdot b_1$ that k has the value $1/V$.

(e) Determine analogous expressions for b_2 and b_3, given that these vectors lie in the directions of $a_3 \times a_1$, and $a_1 \times a_2$, respectively.

(f) Demonstrate that $|b_1 \cdot (b_2 \times b_3)| = 1/V$.

(g) Express the solution of part (c) in the matrix form $b = eX$, where $b = (b_1 \ b_2 \ b_3)$ and $e = (\hat{i} \ \hat{j} \ \hat{k})$, and give the form of X; similarly, give the form of Y where $a = (a_1 \ a_2 \ a_3) = eY$.

(h) Show that $b = aY^{-1}X$.

(i) Determine Y^{-1}, and hence express the b_i in terms of the a_i.

Problem 15.11 shows that the vectors b_1, b_2, b_3 generate a lattice for which the unit cell is of volume $1/V$, where V is the volume of the cell generated by a_1, a_2, a_3. The new lattice is termed the *reciprocal lattice*, in which a general lattice point is given by $K = hb_1 + kb_2 + lb_3$, where h, k, l are the Miller indices (integers) used to label planes in the direct lattice generated by the base vectors a_1, a_2, a_3. K is perpendicular to the crystal plane with Miller indices (hkl).

In Problem 8.7 the structure factor $F(hkl)$ was written in the form

$$F(hkl) = \sum_j^{\text{cell}} f_j e^{2\pi i(hx_j + ky_j + lz_j)} .$$

▷ Notice that an arbitrary point in the reciprocal lattice is given by $K = pb_1 + qb_2 + rb_3$, where p, q, r are real numbers.

Since any arbitrary point in the direct lattice ▷ may be given by $r_j = x_j a_1 + y_j a_2 + z_j a_3$, Equation (8.2) may be written in the more compact form

$$F(hkl) = \sum_j^{\text{cell}} f_j e^{2\pi i K \cdot r_j}$$

using the vector notation. In practice, the 2π factor is sometimes absorbed into the definition of K.

In Problem 15.4, an example is given of Cartesian coordinate data which define the nuclear positions in an isolated molecule in terms of the Cartesian unit vectors with initial points at some chosen origin. In practice, however, a considerable amount of data comes from crystallographic studies and here the atomic position coordinates are given relative to the unit cell base vectors a_1, a_2, a_3 (which are not, in general, orthogonal). It is then a relatively simple procedure to generate bond lengths and interbond angles from these data.

15.6   Differentiation revisited

In our earlier discussion of partial differentiation, the question was posed, but not answered, as to how a partial derivative may be obtained for an arbitrary direction, rather than for the directions associated with the coordinate axes. We are now in a position to answer this question.

Consider the differential of the function $z = f(x, y)$,

$$dz = \frac{\partial z}{\partial x} dx + \frac{\partial z}{\partial y} dy .$$

If we now introduce the plane polar coordinates r, θ such that $x = r \cos \theta$, $y = r \sin \theta$, then the chain rule provides us with a link between the derivatives in the directions of changing x, y and the directions of changing r, θ:

$$dz = \frac{\partial z}{\partial x}(\cos \theta \, dr - r \sin \theta \, d\theta) + \frac{\partial z}{\partial y}(\sin \theta \, dr + r \cos \theta \, d\theta)$$

$$= \left(\frac{\partial z}{\partial x} \cos \theta + \frac{\partial z}{\partial y} \sin \theta\right) dr + \left(-r \frac{\partial z}{\partial x} \sin \theta + r \frac{\partial z}{\partial y} \cos \theta\right) d\theta$$

$$= \frac{\partial z}{\partial r} dr + \frac{\partial z}{\partial \theta} d\theta ,$$

where $\partial z / \partial r$ is the partial derivative of z in a direction making an angle θ with the x-axis .

▷
v is usually written as ∇z and pronounced 'grad z'.

Problem 15.12

If the unit vector directed along the line making an angle θ with the x-axis is denoted by $\hat{r} = \hat{i} \cos \theta + \hat{j} \sin \theta$, and v is defined ▷ according to

$$v = \hat{i} \frac{\partial z}{\partial x} + \hat{j} \frac{\partial z}{\partial y}$$

(a) show that $\dfrac{\partial z}{\partial r} = v \cdot \hat{r}$ and $\dfrac{1}{r}\dfrac{\partial z}{\partial \theta} = v \cdot \hat{k} \times \hat{r}$, where the second

quantity is, in fact, the gradient in a direction perpendicular to $\hat{r}$;

(b) for the function $z = x^2 - 2xy$, show that

$$\frac{\partial z}{\partial r} = 2r \cos \theta(\cos \theta - 2 \sin \theta) , \text{ and, for given } r \neq 0, \text{ find the values}$$

of θ for which this derivative (i) is zero, (ii) attains its maximum value ▷;

(c) deduce that the unit vector in the direction orthogonal to $\hat{r}$ is of the form $\hat{\theta} = \hat{k} \times \hat{r} = -\hat{i} \sin \theta + \hat{j} \cos \theta$.

▷
(0,0) cannot correspond to a maximum or minimum, as the derivative does not vanish in all directions.

15.7   Integration revisited

In Chapter 11, the integration of a function of two or more variables is described for choices of Cartesian and plane polar coordinates, the particular choice depending upon the shape of the region over which the integration is carried out. The main problem in formulating the integration procedure when plane polar coordinates are used relates to the analogue of $dx\,dy$, which is seen to be $r\,dr\,d\theta$ from an examination of the area that is generated when the two coordinates are incremented by the appropriate differential (Figure 11.4). The problem of working out how the element of area is formulated can be approached by considering the transformation between the differentials of the Cartesian and plane polar coordinate systems, using the chain rule as described in Chapter 10. Thus, since x and y are functions of r and θ, we can write the equations

$$dx = \frac{\partial x}{\partial r}\,dr + \frac{\partial x}{\partial \theta}\,d\theta \quad \text{and} \quad dy = \frac{\partial y}{\partial r}\,dr + \frac{\partial y}{\partial \theta}\,d\theta$$

which may be collected into the matrix form

$$\begin{pmatrix} dx \\ dy \end{pmatrix} = \begin{pmatrix} \dfrac{\partial x}{\partial r} & \dfrac{\partial x}{\partial \theta} \\ \dfrac{\partial y}{\partial r} & \dfrac{\partial y}{\partial \theta} \end{pmatrix} \begin{pmatrix} dr \\ d\theta \end{pmatrix} = \begin{pmatrix} \cos\theta & -r\sin\theta \\ \sin\theta & r\cos\theta \end{pmatrix} \begin{pmatrix} dr \\ d\theta \end{pmatrix} \Rightarrow d\mathbf{x} = \mathbf{J}d\mathbf{r}$$

in a matrix notation. $\mathbf{J}$ is termed the Jacobian matrix.

The link between the two elements of area is provided by $\det\mathbf{J}$, the Jacobian for the transformation, in that $dx\,dy = dr\,d\theta\,\det\mathbf{J} = r\,dr\,d\theta$. The important point about writing the chain rule in matrix notation is that the reverse transformation from plane polar to Cartesian coordinates, written as

$$\begin{pmatrix} dr \\ d\theta \end{pmatrix} = \begin{pmatrix} \dfrac{\partial r}{\partial x} & \dfrac{\partial r}{\partial y} \\ \dfrac{\partial \theta}{\partial x} & \dfrac{\partial \theta}{\partial y} \end{pmatrix} \begin{pmatrix} dx \\ dy \end{pmatrix} \Rightarrow d\mathbf{r} = \mathbf{M}d\mathbf{x},$$

enables the partial derivatives of the polar variables with respect to Cartesian variables to be obtained by inverting the matrix $\mathbf{J}$:

$$d\mathbf{x} = \mathbf{J}d\mathbf{r} \Rightarrow \mathbf{J}^{-1}d\mathbf{x} = d\mathbf{r} \Rightarrow \mathbf{M} = \mathbf{J}^{-1}.$$

Problem 15.13

Use the Jacobian matrix $\mathbf{J}$ above for the transformation between Cartesian and plane polar coordinates to determine $\dfrac{\partial r}{\partial x}, \dfrac{\partial r}{\partial y}, \dfrac{\partial \theta}{\partial x}, \dfrac{\partial \theta}{\partial y}$; give answers in both plane polar coordinate and Cartesian coordinate forms.

▷
For a sphere of given radius (the earth), any point on the surface is determined by giving the latitude and longitude.

Problem 15.14

In three dimensions, the Cartesian and spherical polar coordinates are x, y, z and r, θ, ϕ, where r is the radius of the sphere (ranging from 0 to ∞), θ the latitude measured from the z-axis (ranging from 0 to π), and ϕ the longitude measured from the x axis (ranging from 0 to 2π) ▷.

(a) Use the chain rule in the form $dx = \dfrac{\partial x}{\partial r}\, dr + \dfrac{\partial x}{\partial \theta}\, d\theta + \dfrac{\partial x}{\partial \phi}\, d\phi$, etc.,

to write down the Jacobian matrix in terms of r, θ and ϕ, given that $x = r \sin \theta \cos \phi$, $y = r \sin \theta \sin \phi$, and $z = r \cos \theta$

(b) Evaluate $\det \mathbf{J}$ and show that the volume of the small parallelepiped generated when the three polar coordinates are incremented by their differentials is $r^2 \sin \theta dr\, d\theta\, d\phi$.

(c) For the $2p_z$ hydrogen atomic orbital function

$$\psi(r, \theta, \phi) = \left(\frac{1}{32\pi a_0^5}\right)^{\frac{1}{2}} r\, e^{-\frac{r}{2na_0}} \cos \theta,$$

evaluate the integral

$$\int_0^\infty r^2 dr \int_0^\pi \sin \theta\, d\theta \int_0^{2\pi} d\phi\, \psi(r, \theta, \phi)\, r\, \psi(r, \theta, \phi),$$

which corresponds to the average radial distance of the electron from the nucleus.

Summary: This concludes our discussion of the properties and uses of vectors for the time being. The revisiting of the calculus of functions of two or more variables, through the use of matrices, provides an important development that enables us to transform integrals over one set of coordinates to another set through the use of the Jacobian matrix, a necessary prerequisite when modelling the dynamic interactions of molecular species or dealing with the problem of separating the rotational and vibrational motions of an isolated molecule from the corresponding motions of the electrons.

We shall develop the subject matter of this and the previous two chapters in the next chapter on the eigenvalue problem, a central topic in the detailed theoretical understanding of spectroscopy and bonding theories.

16 The eigenvalue problem – an important link between theory and experiment

Objectives

This chapter

- describes intuitively what is meant by an eigenvalue problem

- demonstrates how the previously described solution of homogeneous equations may be used to determine eigenvalues and eigenvectors

- presents examples within a chemical context

The eigenvalue problem provides one of the most important methodologies for quantifying the modelling of chemical behaviour in order to obtain data which are related to experimental observables. The reason that this is the case in chemistry lies at the heart of quantum theory, and any attempt to model molecular structure necessarily involves optimizing the parameters defining the model – and this leads to an eigenvalue problem. In other areas of science and engineering, the models used in simulations use essentially the same mathematics.

16.1 Examples of eigenvalue problems

At the microscopic level in chemistry we might be interested in modelling the vibrations of a polymer molecule. Since we observe the consequences of molecular vibrations in terms of energy absorption or emission involving a transition from one state to another, any model that we use for simulating the effects of the electrons (they hold the atoms together with varying degrees of stiffness) on the vibrations of the nuclei must, if it is to be any good, lead to the best descriptions of the energies of the two states between which the transition is occurring. It is the requirement of seeking the *best* descriptions of the energy states that leads directly to an eigenvalue problem: put another way, out of all the possible motions of the nuclei in a non-linear molecule, only $3N - 6$ modes with characteristic frequencies can arise in practice. The eigenvalue problem in this context arises when we search for the optimum decsription of the nuclear motions because then we discover that only certain energy solutions are permitted. The characteristic values of the energy are termed eigenvalues. Similarly, in electronic structure problems for molecules, when we try to determine the best molecular orbitals for describing the motion of the electrons, we are led to a discrete set of orbitals with characteristic energies (eigenvalues) for describing the energy states of individual electrons.

In a completely different context, standing waves of characteristically different frequencies set up on a stringed musical instrument can be modelled theoretically in terms of the properties of the string to yield eigenvalues corresponding to frequencies of sound.

In a more mundane mathematical situation, the general equation of an ellipse in the xy plane is $Ax^2 + By^2 + 2Cxy = 1$ for an arbitrary choice of axes. If, however, we align the Cartesian axes along the major and minor axes $\triangleright$ of the ellipse (the directions of maximum and minimum fatness), then the equation asssumes the simplest form possible: $Ux^2 + Vy^2 = 1$, in which the values of U and V *alone* characterize the ellipse. The route in going from the general equation to the specific equation requires the solution of an eigenvalue problem to determine the principal axes.

$\triangleright$
Usually termed the
principal axes.

16.2   Defining an eigenvalue problem

Consider the operation of cranking a vector r through an angle θ in the anticlockwise sense (see Figure 16.1). Mathematically we write $\hat{R}\,r = r'$, where $\hat{R}$ is such that $|r| = |r'|$. Now as we have seen in Chapter 15, we can

Figure 16.1
The cranking of the vector r through an angle θ about an axis perpendicular to the paper (viewed down the axis of rotation)

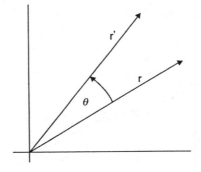

▷
That is, isomorphically;
sometimes, however, we
say that vectors are
mapped onto column
matrices.

associate the vector r in a 1:1 sense ▷ with the column matrix **r**, whose elements are the components of r; similarly, the geometrical operation $\hat{R}$ is linked in a 1:1 manner with the square matrix **R**, whose elements determine how the components of r are related to those of r '. Thus, using the language of matrices, the act of cranking the vector r can be described by the matrix equation **Rr** = **r**'.

For the special kind of operation that only stretches vectors, but with no associated rotation, **r**' remains in the direction of **r**, and hence may be written in the form λ**r**, where λ is a constant. In this situation

$$\mathbf{Rr} = \lambda\mathbf{r}, \tag{16.1}$$

and we have an *eigenvalue* equation, where λ is the eigenvalue and **r** the matrix representative of the *eigenvector*.

16.3   Solving the eigenvalue problem

The question now arises as to how the eigenvalues (λ) and eigenvector representatives (**r**) are determined for a given matrix **R**. It turns out that there may be a number of eigenvalues with characteristic λ values; however, this number cannot exceed n, the order of the matrix **R**.

If Equation (16.1) is written in the equivalent form

$$\mathbf{Rr} = \lambda\mathbf{Er} \Rightarrow (\mathbf{R} - \mathbf{E}\lambda)\mathbf{r} = \mathbf{0}, \tag{16.2}$$

by introducing the unit matrix, **E**, then Equation (16.2) just represents a set of homogeneous equations. As seen in Chapter 10, however, such a set of equations has a non-trivial solution only if $\det(\mathbf{R} - \mathbf{E}\lambda) = 0$. The expansion of the determinant on the left-hand side of this equation yields a polynomial equation of degree n to solve for the permitted values of λ. The eigenvector representative, **r**, associated with each value of λ is then determined by solving the set of homogeneous equations given in Equation (16.2).

Worked example

▷
The eigenvector
representative is often
termed simply the
eigenvector.

16.1 Find the eigenvalues and eigenvector representatives ▷ of matrix **R** given by

$$\mathbf{R} = \begin{pmatrix} 2 & -1 \\ -1 & 2 \end{pmatrix}.$$

Solution $\det(\mathbf{R} - \lambda\mathbf{E}) = 0$

$$\Rightarrow \begin{vmatrix} 2-\lambda & -1 \\ -1 & 2-\lambda \end{vmatrix} = 0 \Rightarrow (2-\lambda)^2 - 1 = 0$$

$$\Rightarrow \lambda^2 - 4\lambda + 3 = 0 \Rightarrow \lambda = 1, \lambda = 3.$$

From Equation (16.2), the eigenvector for $\lambda = 1$ requires

$$\begin{pmatrix} 1 & -1 \\ -1 & 1 \end{pmatrix}\begin{pmatrix} x \\ y \end{pmatrix} = \begin{pmatrix} 0 \\ 0 \end{pmatrix} \Rightarrow x = y \Rightarrow \mathbf{r}_1 = x\begin{pmatrix} 1 \\ 1 \end{pmatrix}.$$

$\triangleright$

This is always the case: for an $n \times n$ matrix only $n - 1$ secular equations are useful.

It should be noticed that although there are two equations, they both lead to the same information $\triangleright$. Similarly, the secular equations for $\lambda = 3$ yield the eigenvector $\mathbf{r}_2 = x\begin{pmatrix} 1 \\ -1 \end{pmatrix}$.

It is usual to choose the value of x such that the vector isomorphic to each matrix representative of an eigenvector is normalized. As seen in Chapter 14, this requires $\mathbf{r}_1^T\mathbf{r}_1 = 1$ and $\mathbf{r}_2^T\mathbf{r}_2 = 1$: that is $2x^2 = 1$ for both eigenvectors, which implies that $x = \pm 1/\sqrt{2}$. It is conventional to select the positive root in each case. The two eigenvector representatives are therefore given by

$$\mathbf{r}_1 = \begin{pmatrix} 1/\sqrt{2} \\ 1/\sqrt{2} \end{pmatrix} \text{ and } \mathbf{r}_2 = \begin{pmatrix} 1/\sqrt{2} \\ -1/\sqrt{2} \end{pmatrix}.$$

Two features arising in the last example are quite general when dealing with symmetric matrices: first, since $\mathbf{r}_1^T\mathbf{r}_2 = 0$, the vector analogue is $\mathbf{r}_1 \cdot \mathbf{r}_2 = 0$ which indicates orthogonality of the eigenvectors. Secondly, if the matrix, $\mathbf{X}$, of eigenvector components is constructed with each eigenvector representative as one column, then the matrix product

$$\mathbf{X}^T\mathbf{R}\mathbf{X} = \begin{pmatrix} 1\sqrt{2} & 1/\sqrt{2} \\ 1/\sqrt{2} & -1/\sqrt{2} \end{pmatrix}\begin{pmatrix} 2 & -1 \\ -1 & 2 \end{pmatrix}\begin{pmatrix} 1/\sqrt{2} & 1/\sqrt{2} \\ 1/\sqrt{2} & -1/\sqrt{2} \end{pmatrix}$$

$$= \begin{pmatrix} 1 & 0 \\ 0 & 3 \end{pmatrix}$$

yields a matrix with the eigenvalues on the diagonal, and zeros everywhere else. We say that the matrix $\mathbf{X}$ diagonalizes $\mathbf{R}$.

In this example, the matrix $\mathbf{X}$ is easily seen to be an orthogonal matrix, as it must be, given that the eigenvectors have been chosen to be orthonormal. Such a choice is always possible when diagonalizing symmetric matrices. When dealing with other types of matrices this result generalizes in different ways. In the case of hermitian matrices, for example, $\mathbf{X}$ can be chosen to be a unitary matrix.

In practice there are several different ways of diagonalizing a given matrix. The simple method described here is satisfactory for matrices of low order; however, for matrices of order four or more it is far more effective to use a computer algorithm which is based on an efficient numerical method for achieving the eigenvalues and eigenvector representatives. If, for one reason or another, two eigenvectors are very nearly the same $\triangleright$, the resulting numerical instability may make it impossible to diagonalize the given matrix. It is often the case that such instability results from an improper formulation of the problem that generates the matrix $\mathbf{R}$. In other cases it might be necessary to use higher precision in the computational procedure $\triangleright$.

$\triangleright$

That is, $\mathbf{r}_1 \cdot \mathbf{r}_2$ is very close to unity.

$\triangleright$

For example, quadruple rather than double precision in FORTRAN programs.

If the diagonalization process yields a set of eigenvalues in which two or more are repeated, then the latter eigenvalues are said to be degenerate. In most cases of interest, the number of eigenvalues (including those that are degenerate) is equal to the order of the matrix which is diagonalized; however, for some matrices, there may be fewer eigenvectors than expected (see, for example, Problem 16.1(b)).

Problem 16.1

Find the eigenvalues and eigenvectors of the following matrices:

(a) $\begin{pmatrix} 5 & 4 \\ 1 & 2 \end{pmatrix}$, (b) $\begin{pmatrix} 1 & 2 \\ 0 & 1 \end{pmatrix}$, (c) $\begin{pmatrix} 13 & -3 & 5 \\ 0 & 4 & 0 \\ -15 & 9 & -7 \end{pmatrix}$

(d) $\begin{pmatrix} 3 & 0 & 0 \\ 0 & 1 & 0 \\ 0 & 0 & 4 \end{pmatrix}$.

16.4   The case of repeated eigenvalues

There are some matrices where the solution of the equations defining the eigenvectors admits of no obviously simple solution. In these cases the difficulty arises because at least two eigenvectors have the same eigenvalue, and their determination requires an additional step which takes account of the fact that different eigenvectors must be orthogonal. In such cases the eigenvectors are not uniquely determined, as any linear combination of eigenvectors also yields an eigenvector; there is thus an element of choice in selecting a set of eigenvectors corresponding to degenerate eigenvalues.

Worked example

16.2   Find the eigenvalues and eigenvectors of the matrix

$$\mathbf{R} = \begin{pmatrix} 1 & 1 & 1 \\ 1 & 1 & 1 \\ 1 & 1 & 1 \end{pmatrix}.$$

Solution   The equation $\det(\mathbf{R} - \lambda\mathbf{E}) = 0$ yields $\lambda^2(\lambda - 3) = 0$, which is satisfied by $\lambda = 0$ and $\lambda = 3$. The eigenvector corresponding to the eigenvalue $\lambda = 3$ is determined in a straightforward manner and has the form

$$\begin{pmatrix} 1/\sqrt{3} \\ 1/\sqrt{3} \\ 1/\sqrt{3} \end{pmatrix} = \frac{1}{\sqrt{3}} \begin{pmatrix} 1 \\ 1 \\ 1 \end{pmatrix}.$$

▷
$x^2 + y^2 + z^2 = 1$ is
required for
normalization.

However, the eigenvalue $\lambda = 0$ yields only the equation $x + y + z = 0$ for determining the eigenvector. Thus, all we know is that $z = -x - y$ ▷, and this enables the eigenvector to be written as a sum of two column matrices (eigenvector representatives):

$$\begin{pmatrix} x \\ y \\ -x-y \end{pmatrix} = \begin{pmatrix} x \\ 0 \\ -x \end{pmatrix} + \begin{pmatrix} 0 \\ y \\ -y \end{pmatrix} = x \begin{pmatrix} 1 \\ 0 \\ -1 \end{pmatrix} + y \begin{pmatrix} 0 \\ 1 \\ -1 \end{pmatrix}$$

$$= x \, \mathbf{r}_1 + y \, \mathbf{r}_2 \, .$$

▷
Eigenvectors with a
common eigenvalue are
said to be degenerate.
▷
A $(1/\sqrt{2})$ factor ensures
normalization.

Although the vectors $x \, \hat{\mathbf{i}} - x \, \hat{\mathbf{k}}$ and $y \, \hat{\mathbf{j}} - y \, \hat{\mathbf{k}}$, corresponding to the two column matrices, are not orthogonal, both column matrices, $\mathbf{r}_i, (i = 1, 2)$, have the property that $\mathbf{R}\mathbf{r}_i = 0$, indicating that the eigenvalue zero is repeated twice ▷. In such a situation, any linear combination of these column matrices of the form $\mathbf{v} = x \, \mathbf{r}_1 + y \, \mathbf{r}_2$ is also associated with the eigenvalue zero, since $\mathbf{R}\mathbf{v} = x \, \mathbf{R}\mathbf{r}_1 + y \, \mathbf{R}\mathbf{r}_2 = 0$. It just remains, therefore, to find two column matrices which are isomorphic to orthogonal vectors, and hence to eigenvectors. Take $\mathbf{r}_2$ as one eigenvector (not normalized) ▷; the other eigenvector, $\mathbf{v}$, is determined by requiring that $\mathbf{r}_2^T \mathbf{v} = 0$. The latter orthogonality constraint results in the equation $x + 2y = 0$. Thus,

$$\mathbf{v} = -2y \, \mathbf{r}_1 + y \, \mathbf{r}_2 = y \begin{pmatrix} -2 \\ 1 \\ 1 \end{pmatrix},$$

which yields a second normalized eigenvector in the form

$$\frac{1}{\sqrt{6}} \begin{pmatrix} -2 \\ 1 \\ 1 \end{pmatrix}.$$

▷
See Problem 14.8.

Problem 16.2

The cyclopropenyl cation, $C_3H_3^+$, yields the following determinantal equation to solve for the molecular orbital energy levels (eigenvalues) ▷:

$$\begin{vmatrix} \alpha - \lambda & \beta & \beta \\ \beta & \alpha - \lambda & \beta \\ \beta & \beta & \alpha - \lambda \end{vmatrix} = 0 \, .$$

Use the method in the previous example to determine the eigenvectors (molecular orbitals, ψ), in which the coefficient of each π-type atomic orbital, ϕ_i, is given by x, y, z for a given molecular orbital of the form $\psi = x\phi_1 + y\phi_2 + z\phi_3$. The atomic orbitals here take the place of the unit base vectors in the earlier discussion, and the column matrix of coefficients x, y, z is now the matrix representative of the molecular orbital, ψ.

The principal axis transformation

In the introductory section of this chapter, it was stated that the general equation for an ellipse in the xy plane could be rewritten in a simpler form by redefining the axes. The procedure involves first characterizing the ellipse by a matrix, and then using the eigenvectors of this matrix to determine the principal axes. The sequence of such steps, as described in detail below, provides a way of determining the preferred axes for describing chemical properties such as electronic polarizability, the dynamics of nuclear motions (as they manifest themselves in rotational spectroscopy), as well as in magnetic susceptibility and electron spin resonance spectroscopy ▷.

▷
Alberty and Silbey,
Sections 14.5, 14.9;
Carrington and
McLachlan, Chapter 7.

Consider again the equation of an ellipse, which is a second-degree polynomial in the two variables x and y: $Ax^2 + By^2 + 2Cxy = 1$. This equation may be written in matrix notation if the x and y coordinates are first collected into a column matrix, $\mathbf{r}$, and the coefficients are then collected into a square 2×2 matrix, $\mathbf{D}$:

$$(x \quad y)\begin{pmatrix} A & C \\ C & B \end{pmatrix}\begin{pmatrix} x \\ y \end{pmatrix} = \mathbf{r}^T\mathbf{D}\mathbf{r} = 1 .$$

▷
D is symmetric, and
hence **X** has the property
XX^T = **E**.

The principal axes are determined by first diagonalizing $\mathbf{D}$, and then forming the (orthogonal) ▷ matrix of eigenvectors. It then follows by inserting the unit matrix in the form $\mathbf{XX}^T$ that

$$\mathbf{r}^T\mathbf{D}\mathbf{r} = \mathbf{r}^T\mathbf{XX}^T\mathbf{D}\mathbf{XX}^T\mathbf{r} = 1 \Rightarrow \mathbf{r}'^T\mathbf{D}_d\mathbf{r}' = 1 ,$$

where $\mathbf{D}_d$ is diagonal and $\mathbf{r}' = \mathbf{X}^T\mathbf{r}$ defines the new coordinates in the transformed axis system in terms of the old coordinates.

▷
Notice that each
eigenvector is
undetermined to within a
sign; a particular choice
is made here.

> ### Problem 16.3
>
> For the ellipse $2x^2 + 3y^2 - 2\sqrt{2}xy = 1$, described in a Cartesian coordinate system with base vectors $\hat{\imath}, \hat{\jmath}$, show that
>
> (a) $\mathbf{D} = \begin{pmatrix} 2 & -\sqrt{2} \\ -\sqrt{2} & 3 \end{pmatrix}$
>
> (b) the eigenvalues of $\mathbf{D}$ are 1 and 4
>
> (c) the matrix of eigenvectors ▷ is
>
> $$\mathbf{X} = \begin{pmatrix} \sqrt{2/3} & \sqrt{1/3} \\ \sqrt{1/3} & -\sqrt{2/3} \end{pmatrix}$$
>
> (d) $\mathbf{r}' = \begin{pmatrix} x' \\ y' \end{pmatrix} = \frac{1}{\sqrt{3}}\begin{pmatrix} x\sqrt{2} + y \\ x - y\sqrt{2} \end{pmatrix}$
>
> (e) the equation of the ellipse in the (rotated) coordinate system, defined with respect to the base vectors $\hat{\imath}', \hat{\jmath}'$, is given by $x'^2 + 4y'^2 = 1$.

In Problem 16.3 it is seen how the transformation of the equation of an ellipse can lead to a simpler form, corresponding to a new choice of axes. The eigenvectors of the matrix defining the ellipse in the original coordinate system may be used to specify how the new (rotated) coordinate system is related to the initial coordinate system, as well as the form of the new base vectors in terms of $\hat{i}, \hat{j}$.

Since the vector from the origin to an arbitrary point on the ellipse can be given in terms of either the new coordinates (x', y'), defined with respect to the base vectors $\hat{i}', \hat{j}'$, or the old coordinates (x, y), defined with respect to $\hat{i}, \hat{j}$:

$$\mathbf{r} = x'\hat{i}' + y'\hat{j}' = x\hat{i} + y\hat{j}, \tag{16.3}$$

it follows from the expression for $\mathbf{X}$ in Problem 16.3(d) that

$$\begin{pmatrix} x' \\ y' \end{pmatrix} = \begin{pmatrix} x\sqrt{2/3} + y\sqrt{1/3} \\ x\sqrt{1/3} - y\sqrt{2/3} \end{pmatrix} \Rightarrow x'\begin{pmatrix} 1 \\ 0 \end{pmatrix} + y'\begin{pmatrix} 0 \\ 1 \end{pmatrix}$$

$$= x\begin{pmatrix} \sqrt{2/3} \\ \sqrt{1/3} \end{pmatrix} + y\begin{pmatrix} \sqrt{1/3} \\ -\sqrt{2/3} \end{pmatrix}.$$

However, on associating column matrices with unit vectors, we have

$$x'\hat{i}' + y'\hat{j}' = x(\sqrt{2/3}\,\hat{i}' + \sqrt{1/3}\,\hat{j}') + y(\sqrt{1/3}\,\hat{i}' - \sqrt{2/3}\,\hat{j}')$$

$$= x\hat{i} + y\hat{j},$$

where the last result follows from Equation (16.3). Hence, $\hat{i} = \sqrt{2/3}\,\hat{i}' + \sqrt{1/3}\,\hat{j}'$, and $\hat{j} = \sqrt{1/3}\,\hat{i}' - \sqrt{2/3}\,\hat{j}'$. If θ_x is the angle between $\hat{i}'$ and $\hat{i}$ then $\hat{i}' \cdot \hat{i} = \sqrt{2/3} = \cos\theta_x \Rightarrow \theta_x = 35.3°$.

Problem 16.4

For the ellipse problem just described above,

(a) take the scalar product $\hat{i}' \cdot \hat{i}$ and thus determine the position of $\hat{i}'$ in the xy plane
(b) use the scalar products $\hat{j}' \cdot \hat{i}$ and $\hat{j}' \cdot \hat{j}$ to determine the orientation of $\hat{j}'$ with respect to the old base vectors
(c) suggest other possible orientations of $\hat{i}'$ and $\hat{j}'$ if different signs for the normalization constants are chosen in the determination of the eigenvectors of $\mathbf{D}$ ▷.

▷
There are four possibilities altogether.

Summary: In this chapter on the eigenvalue problems, we first of all discussed how such problems arise in a chemical context and then concentrated more on their algebraic solution. Practical methods for obtaining eigenvalues and eigenvectors of a square matrix are best obtained through the use of a numerical algorithm – a common step involving the reduction of the initial matrix to tridiagonal form (see, for example, Press *et al.*, 1986). Suitable algorithms are available in libraries on most computer systems. However, as noted in our earlier discussion in this chapter, caution has to be exercised when the matrix is ill-conditioned in the sense that two or more eigenvectors may be very similar in form (that is, nearly identical): the problem is then not with the numerical algorithm, but with the formulation of the model.

In the next, and concluding, chapter we explore another application of matrix methods in discussing simple curve fitting procedures which are required in the processing of experimental data.

17 Curve fitting – vectors revisited

Objectives

This chapter establishes the ideas of

- linear dependence and independence of vectors
- orthogonal projection of a vector in order to define a measure of closeness
- fitting data points to a polynomial function

▷

For example, ln x.

In chemistry it is often necessary to take a given set of data points (x_i, y_i) $(i = 1, 2, 3, \ldots)$ and process them in some way to establish a relation between two sets of variables, x and y. The reason for doing this is often related to the use of a theoretical model which relates x, or some function of x ▷, to y, or some function of y. Because of experimental error, or the restricted application of the theoretical model, the processed data do not necessarily satisfy the equation derived from theory. The problem then arises as to how the best theoretical curve (linear, quadratic, etc.) can be drawn through the set of data points, when there is no guarantee that any of the data points will lie on the curve. This problem is often solved using the tools of calculus within the context of statistics and error analysis (see Chapter 12). A different approach is adopted here in order to bring together some principles developed in earlier chapters on matrices, determinants and vectors. The reason for doing this is to show how some powerful techniques in vector analysis can be deployed which have far more generality than the particular application considered in this chapter.

17.1   Base vectors revisited

In Chapter 15 the concept of a set of base vectors was described and limited to two or three dimensions. Instead of referring to such unit vectors as $\hat{i}, \hat{j}, \hat{k}$, it is convenient to introduce a notation which admits of a simple generalization to any number of dimensions: thus, an arbitrary vector, u, in three-dimensional space is written as $x_1\hat{e}_1 + x_2\hat{e}_2 + x_3\hat{e}_3$, where the components x_1, x_2, x_3 are real numbers. The basis vectors are said to *span* the three-dimensional space – simply because all vectors in the space are generated as the components take on all their respective permitted values. In other words, *any* other vector in three-dimensional space may be written as a sum of multiples of the three base vectors. It should also be noted that the Cartesian unit vectors discussed here are normalized and orthogonal $\triangleright$ in the sense that $\hat{e}_i \cdot \hat{e}_j = \delta_{ij}$.

$\triangleright$
Of unit length and at right angles, respectively.

The choice of orthogonal base vectors is often one of convenience, and not of necessity; but, as seen in Problem 15.11, the unit cell in a crystalline solid may well be described in terms of base vectors which are not orthogonal. However, whatever base vectors are chosen for spanning three-dimensional space, some care is necessary in ensuring that they do not all lie in the same plane. If they are coplanar, on the other hand, then no vector lying in a direction perpendicular to their plane can be formed by linear combination of them, and one of the three so-called base vectors is redundant. When this happens, we say that either the three base vectors are not *linearly independent* or, equivalently, there is *linear dependence*. In practical terms this means that one of the three vectors can be expressed in terms of the other two: hence the notion of linear dependence, and the consequent lack of linear independence. For three such vectors, $\hat{v}_1, \hat{v}_2, \hat{v}_3$, the test for linear dependence (or lack of it) is best carried out using the scalar product of $\hat{v}_1$, say, with the vector $\hat{v}_2 \times \hat{v}_3$. The latter vector is perpendicular to the plane containing $\hat{v}_2$ and $\hat{v}_3$, and so if $\hat{v}_1$ lies in the same plane then

$$\hat{v}_1 \cdot (\hat{v}_2 \times \hat{v}_3) = 0,$$

and linear dependence is present.

Worked example

17.1   Show that the vectors

$$\hat{v}_1 = \frac{1}{\sqrt{2}}(\hat{e}_1 + \hat{e}_2), \quad \hat{v}_2 = \frac{1}{\sqrt{2}}(\hat{e}_1 + \hat{e}_3), \quad \hat{v}_3 = \frac{1}{\sqrt{2}}(\hat{e}_2 - \hat{e}_3)$$

exhibit linear dependence.

Solution
$$\hat{v}_1 \cdot (\hat{v}_2 \times \hat{v}_3) = \begin{vmatrix} 1 & 1 & 0 \\ 1 & 0 & 1 \\ 0 & 1 & -1 \end{vmatrix} = 0,$$

where the determinant may be evaluated either by direct expansion or by first adding the third row to the second row and then noting that the resulting determinant has two identical rows, and thus vanishes. The three vectors are therefore not linearly independent, and one vector may be expressed in terms of the other two: for example, $\hat{v}_3 = \hat{v}_1 - \hat{v}_2$. Expressed another way, $\hat{v}_3$ lies in the space spanned by $\hat{v}_1$ and $\hat{v}_2$.

Dealing with n dimensions

Despite the simplicity of the above test for linear dependence, its use is clearly limited to vectors in three dimensions. For an n-dimensional space, an arbitrary vector may be expressed in terms of a linear combination of the n base vectors:

$$u = x_1\hat{e}_1 + x_2\hat{e}_2 + x_3\hat{e}_3 + \cdots + x_n\hat{e}_n .$$

In some problems it may be more convenient to construct combinations of the base vectors $\hat{e}_i$ to form a new set of base vectors – either spanning the full space or some subspace of the full space. These base vectors, u_i, may or may not be normalized. In all such situations, it is important to have a simple test for linear dependence. The test that is widely used involves constructing the

▷

Usually called the Gram determinant.

determinant ▷ of scalar products of pairs of (new) base vectors, $u_i \cdot u_j$:

$$\det \mathbf{S} = \begin{vmatrix} u_1 \cdot u_1 & u_1 \cdot u_2 & \cdots & u_1 \cdot u_n \\ u_2 \cdot u_1 & u_2 \cdot u_2 & \cdots & u_2 \cdot u_n \\ \vdots & \vdots & \vdots & \vdots \\ u_n \cdot u_1 & u_n \cdot u_2 & \cdots & u_n \cdot u_n \end{vmatrix} = \begin{vmatrix} S_{11} & S_{12} & \cdots & S_{1n} \\ S_{21} & S_{22} & \cdots & S_{2n} \\ \vdots & \vdots & \vdots & \vdots \\ S_{n1} & S_{n2} & \cdots & S_{nn} \end{vmatrix}$$

and verifying that $\det \mathbf{S} \neq 0$.

17.2   Projecting a vector onto a subspace

Consider the three vectors u_1, u_2 and v, where v is not in the plane spanned by u_1 and u_2 (Figure 17.1). From the earlier discussion of vectors, v may be replaced by the sum of two vectors: one, s, orthogonal to the plane containing

▷

s is orthogonal to both u_1 and u_2.

u_1, u_2 ▷, and the other, w, lying in the plane. This latter vector w, termed the

Figure 17.1
The projection of a vector v on to the plane containing the base vectors u_1, u_2; Q lies in the plane

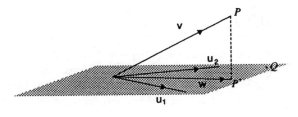

orthogonal projection of v on to the plane (subspace) spanned by u_1 and u_2, may thus be written in the form $w = a_1 u_1 + a_2 u_2$. In fact, w is the 'closest' approximation to v that can be constructed from u_1 and u_2; closest being interpreted in the geometrical sense in terms of the shortest distance from P to the point P' in the plane. All other points, Q, in the plane are such that $|\overrightarrow{PQ}| > |\overrightarrow{PP'}|$ since, from Pythagoras (Figure 17.1), $|\overrightarrow{PQ}|^2 = |\overrightarrow{PP'}|^2 + |\overrightarrow{P'Q}|^2 \Rightarrow |\overrightarrow{PQ}| > |\overrightarrow{PP'}|$, as $|\overrightarrow{P'Q}| > 0$.

Now from the rule for the addition of vectors, $v = w + s$, and $s = v - w$ is orthogonal to u_1 and u_2 by construction. It follows, therefore, that

$$u_1 \cdot (v - w) = 0 \qquad \text{and} \qquad u_2 \cdot (v - w) = 0,$$

and substitution for w in terms of u_1 and u_2 yields the equations

$$v_1 - a_1 S_{11} - a_2 S_{12} = 0 \qquad \text{and} \qquad v_2 - a_1 S_{21} - a_2 S_{22} = 0,$$

$$(17.1)$$

where $S_{ij} = u_i \cdot u_j$, and $v_i = u_i \cdot v$. Equations (17.1) may be written in the matrix form

$$\begin{pmatrix} v_1 \\ v_2 \end{pmatrix} = \begin{pmatrix} S_{11} & S_{12} \\ S_{21} & S_{22} \end{pmatrix} \begin{pmatrix} a_1 \\ a_2 \end{pmatrix} \Rightarrow \begin{pmatrix} a_1 \\ a_2 \end{pmatrix} = \begin{pmatrix} S_{11} & S_{12} \\ S_{21} & S_{22} \end{pmatrix}^{-1} \begin{pmatrix} v_1 \\ v_2 \end{pmatrix},$$

$$(17.2)$$

▷

The *metrical* matrix.

thus demonstrating that the best approximation to the vector v by the vector w, in the subspace spanned by u_1 and u_2, involves inverting the matrix **S** ▷ associated with the subspace of interest, prior to calculating the best values for the coefficients a_1, a_2..

Problem 17.1

Determine the orthogonal projection, w, of the following vectors, v, onto the subspaces spanned by the given vectors, u_i $(i = 1, 2)$:

(a) $v = x_1 \hat{e}_1 + x_2 \hat{e}_2 + x_3 \hat{e}_3$, for $u_1 = \hat{e}_1$, and $u_2 = \hat{e}_1 + \hat{e}_2 + \hat{e}_3$

(b) $v = \hat{e}_1 + \hat{e}_3$, for $u_1 = \hat{e}_2 + \hat{e}_3$, and $u_2 = \hat{e}_1 + \hat{e}_2$.

17.3   Curve fitting

The straight line

As we have seen in Chapters 2 and 12, it is common in chemistry for the model describing a physical process to yield a (decaying) exponential dependence of concentration of a species on, for example, time (in a chemical reaction) or on distance (in absorption spectroscopy). In the former situation,

▷
See Problem 2.10.

where it is necessary to extract the rate constant in a kinetic experiment, it is technically simpler to take the logarithm of the concentration as the dependent variable, so that a linear relation may be plotted ▷. The slope of the resulting straight line is then proportional to the rate constant. Another more compelling reason for linearizing the data is that, because of experimental errors or deficiencies in the model used, all the points will usually not lie on the theoretically determined straight line. In this situation, it becomes much easier to determine the best straight line, with associated estimates for the errors in both the slope and the intercept on each axis.

Before dealing with the general straight line, we shall consider the slightly simpler situation where one property, y, is directly proportional to another property, x, so that the linear relation is $y = a_1 x$, where a_1 is the slope of the line which is supposed to pass through the origin. Consider the set of three points $\{(x_i, y_i)\} = \{(1, 1.2), (2, 2.8), (3, 3.2)\}$ which are obtained from an experiment in which the property described by the variable x (say time) has the values $x_1 = 1$, $x_2 = 2$, $x_3 = 3$. Corresponding to each x_i the value of the property represented by y is y_i (say the logarithm of the concentration at time x_i).

If a straight line relation is sought between x and y in the form $y = mx$, then it remains to determine the value of m such that the straight line fits the data in the best way. The definition of 'best' needs further clarification but, if we can pose the problem in vector form, then we can use the ideas of closeness discussed in the previous section on forming the orthogonal projection of a vector. In particular, the best approximation to a vector v, which is not contained fully within the subspace spanned by $u_1, u_2, \ldots, u_n$, is obtained in terms of its orthogonal projection onto the n-dimensional subspace. The question here is: how are the vectors v, w, $u_1, u_2, \ldots, u_n$ defined?

In the present example, where there are three data points, then a straight line of the form $y = mx$ requires the determination of a value for m such that mx_1, mx_2 and mx_3, are as close as possible to the given y values. We saw in Chapter 12 that the sum of the squares of the errors $y_i - mx_i$ could be minimized by applying the methods of calculus. Here, we use a procedure (with the same outcome) that is based on the use of vectors.

▷
Remember (Chapter 15)
that $u_1 \cdot v = \mathbf{u}_1^T \mathbf{v}$, etc.

If the x components and the y components of the three data points are collected into the column matrix representatives ▷

$$\mathbf{u}_1 = \begin{pmatrix} 1 \\ 2 \\ 3 \end{pmatrix}, \qquad \mathbf{v} = \begin{pmatrix} 1.2 \\ 2.8 \\ 3.2 \end{pmatrix},$$

respectively, then the the subspace onto which v must be projected is one-dimensional, and $w = a_1 u_1$ is the best approximation to v. However, the simple adaptation of Equation (17.1) to one base vector yields $v_1 = a_1 S_{11}$; that is, $a_1 = v_1/S_{11} = 16.4/14 = 1.17$. Thus the y values for the points on the best straight line, with slope $m = a_1 = 1.17$, are 1.17, 2.34 and 3.51, and it is clear that none of the observed data points lie on this line.

If there are n data points initially, then u_1, v and w are all $n \times 1$ matrices.

The general straight line

The analysis given in the last section may be extended readily to determine the best fit of a set of points to a line in the form $y = a_2 + a_1 x$. Since the coefficient of the constant term a_2 is 1, it is convenient to define a vector $\mathbf{u}_2$ according to

$$\mathbf{u}_2 = \begin{pmatrix} 1 \\ 1 \\ 1 \end{pmatrix}$$

▷

Represented by the column matrix of y_i values.

Thus the best approximation to the vector $\mathbf{v}$ ▷ is of the form $\mathbf{w} = a_2 \mathbf{u}_2 + a_1 \mathbf{u}_1$, where the values of a_1 and a_2 are obtained directly from Equation (17.2)

Problem 17.2

Show that for n data points (x_i, y_i), $(i = 1, 2, \ldots, n)$,

(a) $S_{11} = \mathbf{u}_1 \cdot \mathbf{u}_1 = \mathbf{u}_1^T \mathbf{u}_1 = \sum_{i=1}^{n} x_i^2$,

$S_{12} = \mathbf{u}_1 \cdot \mathbf{u}_2 = \mathbf{u}_1^T \mathbf{u}_2 = S_{21} = \sum_{i=1}^{n} x_i$,

$S_{22} = \mathbf{u}_2 \cdot \mathbf{u}_2 = \mathbf{u}_2^T \mathbf{u}_2 = \sum_{i=1}^{n} 1 = n$

(b) $\mathbf{u}_1 \cdot \mathbf{v} = \mathbf{u}_1^T \mathbf{v} = v_1 = \sum_{i=1}^{n} x_i y_i$, $\qquad \mathbf{u}_2 \cdot \mathbf{v} = \mathbf{u}_2^T \mathbf{v} = v_2 \sum_{i=1}^{n} y_i$

(c) $a_2 = \dfrac{\left(\sum_{i=1}^{n} x_i^2\right)\left(\sum_{i=1}^{n} y_i\right) - \left(\sum_{i=1}^{n} x_i\right)\left(\sum_{i=1}^{n} x_i y_i\right)}{n\left(\sum_{i=1}^{n} x_i^2\right) - \left(\sum_{i=1}^{n} x_i\right)^2}$

$a_1 = \dfrac{n\left(\sum_{i=1}^{n} x_i y_i\right) - \left(\sum_{i=1}^{n} x_i\right)\left(\sum_{i=1}^{n} y_i\right)}{n\left(\sum_{i=1}^{n} x_i^2\right) - \left(\sum_{i=1}^{n} x_i\right)^2}$.

The values of a_1 and a_2 obtained in the last problem are the same as those derived in Equations (12.2), (12.3), after substituting for $\bar{x}$ and $\bar{y}$.

Fitting a second-degree polynomial

In some instances the experimental data points are fitted to a polynomial of degree two in the form $y = a_2 + a_1 x + a_3 x^2$. The problem of determining the optimum values of the a_i is now one of taking the orthogonal projection of the vector

$$\mathbf{v} = y_1 \hat{\mathbf{e}}_1 + y_2 \hat{\mathbf{e}}_2 + \cdots + y_n \hat{\mathbf{e}}_n$$

onto the three-dimensional subspace spanned by the vectors

$$u_1 = x_1 \hat{e}_1 + x_2 \hat{e}_2 + \cdots + x_n \hat{e}_n ,$$
$$u_2 = \hat{e}_1 + \hat{e}_2 + \cdots + \hat{e}_n ,$$
$$u_3 = x_1^2 \hat{e}_1 + x_2^2 \hat{e}_2 + \cdots + x_n^2 \hat{e}_n .$$

Clearly, the method described here can be extended to a polynomial function of degree m: the determination of the $m + 1$ linear coefficients specifying the polynomial then involves inverting the matrix S with degree $m + 1$. In practice, however, it is sometimes more advantageous to fit data points with other forms of function – for example, cubic splines or Tschebychev polynomials, both of which have special algebraic and numerical properties.

17.4 Conclusion

This chapter on fitting procedures provides an appropriate point to conclude our journey in which we have examined the deployment of a number of mathematical techniques within the context of chemistry. In this single important application (and the eigenvalue problem discussed in the last chapter), we see another problem solved using the tools of vectors, matrices and determinants.

As we stated in the Introduction, it has not been our objective to be exhaustive in the treatment of mathematical topics; rather, by being selective in singling out important areas of calculus and linear algebra, preparation is made for further explorations of related mathematical tools should the need arise. For example, most of the knowledge required for studying functions of a complex variable, or for applying transform techniques in the solution of differential equations, or for deriving new forms of series that appear in the advanced study of X-ray diffraction, is accessible from the material presented in this introductory text.

References

M Abramowitz and I A Stegun, Editors, *Handbook of Mathematical Functions, with Formulae, Graphs and Mathematical Tables*, Dover, New York, 1965.

R A Alberty and R J Silbey, *Physical Chemistry*, John Wiley and Sons, Inc., New York, 1992.

P W Atkins, *Physical Chemistry*, 5th edition, Oxford University Press, 1994.

G M Barrow, *Physical Chemistry*, 5th edition, McGraw-Hill Book Company, New York, 1988.

R S Berry, S A Rice and J Ross, *Physical Chemistry*, John Wiley and Sons, New York, 1980.

W H Beyer, *CRC Standard Mathematical Tables*, 25th edition, CRC Press Inc., West Palm Beach, 1973.

D M Bishop, *Group Theory for Chemists*, Clarendon Press, Oxford, 1973.

A Carrington and A D McLachlan, *Introduction to Magnetic Resonance*, Chapter 7, Harper and Row, New York, 1967.

Chemical Abstracts, 12th Collected Index, Volumes 106–115, American Chemical Society, 1992.

C A Coulson, *Bull. Inst. Mathematics and its Applications*, **9**, 206, 1973.

A Croft and R R Davison, *Foundation Maths*, Longman, 1995.

O L Davies and P L Goldsmith (Editors), *Statistical Methods in Research and Production*, 4th edition, Oliver and Boyd, Edinburgh, 1972.

J A Duffy, *Bonding, Energy Levels and Bonds in Inorganic Solids*, Longman, 1990.

H B Dwight, *Tables of integrals and other mathematical data*, 4th edition, Macmillan, 1961.

L Goldberg, *Physics Today*, **41**, 38, 1988.

I S Gradshteyn and I M Ryzhik, *Tables of Integrals, Series, and Products*, 5th edition, Ed. A Jeffrey, Academic Press Inc., New York, 1993.

N N Greenwood and A Earnshaw, *Chemistry of the Elements*, Pergamon Press, Oxford, 1984.

I Gutman and O E Polansky, *Mathematical Concepts in Organic Chemistry*, Springer-Verlag, Berlin, 1993.

C N Hinshelwood and R E Burk, *J. Chem. Soc.*, **127**, 1114, 1925.

E R Lapwood, *Ordinary Differential Equations*, Topic 1, Volume 1 of *The International Encyclopedia of Physical Chemistry and Chemical Physics*, Pergamon Press, Oxford, 1968.

P M Lee, *Bayesian Statistics*, Oxford University Press, New York, 1989.

I N Levine, *Quantum Chemistry*, 4th edition, Allyn Bacon Press, Boston, 1991.

G Matthews, *Calculus*, John Murray, London, 1980.

I Mills, T Cvitaš, K Homann, N Kallay and K Kuchitsu, *Quantities, Units and Symbols in Physical Chemistry*, Blackwell, Oxford, 1989.

W H Press, B P Flannery, S A Tenkolsky and W T Vetterling, *Numerical Recipes*, Chapter 11, Cambridge University Press, 1986.

J R Raley, F F Rust and W E Vaughan, *Journal of the American Chemical Society*, **70**, 88, 1948.

S L Salas and E Hille, *Calculus — one and several variables*, 6th edition, John Wiley and Sons, New York, 1990.

D F Shriver, P W Atkins and C H Langford, *Inorganic Chemistry*, 2nd edition, Oxford University Press, 1994.

L Smart and E Moore, *Solid State Chemistry*, Chapman and Hall, London, 1992.

G W Snecedor and W G Cochran, *Statistical Methods*, 7th edition, The Iowa State University Press, Ames, Iowa, 1980.

G Stephenson, *Mathematical Methods for Science Students*, 2nd edition, Longman, London, 1973.

A H Stroud and Don Secrest, *Gaussian Quadrature Formulas*, Prentice-Hall, Inc., Englewood Cliffs, 1966.

S Taylor and D Williams, *Chemistry in Britain*, **29**, 680, 1993.

A Vincent, *Molecular Symmetry and Group Theory*, John Wiley and Sons, Chichester, 1977.

R J Wonnacott and T H Wonnacott, *STATISTICS: Discovering Its Power*, John Wiley and Sons, Inc., New York, 1982.

R H Woodward and R Hoffmann, *The Conservation of Orbital Symmetry*, Academic Press, New York, 1970.

Appendix **1**

SI prefixes and symbols for *n*th powers of 10

n	12	9	6	3	2	1	−1	−2	−3	−6	−9	−12	−15	−18
Prefix	tera	giga	mega	kilo	hecto	deca	deci	centi	milli	micro	nano	pico	femto	atto
Symbol	T	G	M	k	h	da	d	c	m	μ	n	p	f	a

Appendix **2**

Some trigonometry

The sine addition formula

Consider the triangle PQR with sides of length $a + b$, c, d and internal angles $A + B$, C, D as indicated in Figure A2.1. The area, X, of this triangle with base $r = (a + b)$ is given by $X = \frac{1}{2}r \times h$ ▷. However, if the base is taken as d, then the appropriate height is the (external) line TR of length m: a result which follows from the observation that the area, Y, of the rectangle $PTRV$ is $m(d + x)$, where x is the length of QT; but Y is the sum of the areas of the three triangles PQR, RQT, and PRV ▷:

▷

Half base times vertical height, h.

▷

The area of this triangle is half the area of the rectangle $PTRV$.

$$Y = X + \frac{1}{2}mx + \frac{1}{2}m(d + x) = m(d + x) \Rightarrow X = \frac{1}{2}md. \qquad \text{(A2.1)}$$

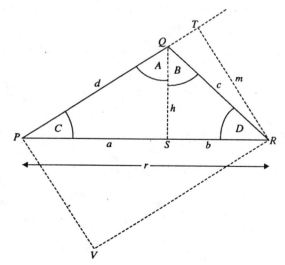

Figure A2.1
A construction based on the triangle PQR, with perpendiculars dropped from Q to S on PR and from R to T on the extension of the line PQ; the point V defines the fourth vertex of the rectangle $PTRV$

Now as the sum of the interior angles of a triangle is 180°, it follows that $(A + B) + C + D = 180°$ and $R\widehat{Q}T = 180° - (A + B) = (C + D)$. Thus, $\sin(C + D) = m/c$ and

$$\sin C = h/d, \sin D = h/c, \cos C = a/d, \text{ and } \cos D = b/c. \qquad \text{(A2.2)}$$

$\triangleright$

h is expressed in terms of sin D using Equation (A2.2).

The equality of the two expressions for the area X implies, therefore, that $\triangleright$

$$X = \frac{1}{2}rh = \frac{1}{2}(ah + bh) = \frac{1}{2}(ac \sin D + bc \sin D) \text{ and}$$

$$X = \frac{1}{2}md = \frac{1}{2}cd \sin(C + D). \qquad \text{(A2.3)}$$

Thus, on equating the two expressions for X and cancelling the common factors of $\frac{1}{2}$ and c, the following identity is obtained:

$$d \sin(C + D) = a \sin D + b \sin D. \qquad \text{(A2.4)}$$

$\triangleright$

$\frac{c}{d} = \frac{c}{h} \cdot \frac{h}{d} = \frac{\sin C}{\sin D}.$

Substituting for a and b from the last two equalities in Equation (A2.2), then dividing through by d, and eliminating c/d $\triangleright$ yields

$$\begin{aligned}\sin(C + D) &= \sin D \cos C + \cos D \sin C \\ &= \sin C \cos D + \cos C \sin D\end{aligned} \qquad \text{(A2.5)}$$

Results derived from the sine addition formula

A number of important results are derived from Equation (A2.5) by making suitable choices for C and D, and remembering how the function $\sin D$ is defined (Figure 2.4, where θ now stands for the angle C or D as appropriate).

$\triangleright$

In these cases, the values of x/r and y/r for a point P on the circle in Figure 2.4 have values of $(-1)^n$ and 0, respectively.

- If D is a multiple of π ($D = n\pi$, where $n = 0, \pm1, \pm2, \ldots$ $\triangleright$) then $\sin D = 0$, $\cos D = (-1)^n$ and

$$\sin(C + n\pi) = (-1)^n \sin C. \qquad \text{(A2.6)}$$

$\triangleright$

The values of y/r and x/r for the point P are $(-1)^n$ and 0, respectively.

- Similarly, if D is an odd multiple of $\pi/2$ ($D = (2n + 1)\pi/2$), where $n = 0, 1, 2, \ldots$ $\triangleright$ then $\sin D = (-1)^n$, $\cos D = 0$, and

$$\sin(C + (2n + 1)\pi/2) = (-1)^n \cos C. \qquad \text{(A2.7)}$$

- If $D = C$, then

$$\sin 2C = 2 \sin C \cos C. \qquad \text{(A2.8)}$$

More properties of the sine function

Substitution of $C = 2\pi$, $D = -C$ in Equation (A2.5) yields

$$\sin(2\pi - C) = \sin(-C); \qquad \text{(A2.9)}$$

but, from Figure 2.4, the value of y/r associated with the angle $(2\pi - C)$ (defined in a clockwise manner from the x-axis), is equal to the negative of the value associated with C, and hence

$$\sin(-C) = -\sin C. \tag{A2.10}$$

▷

In general $f(x)$ is odd if $f(x) = -f(-x)$, and even if $f(x) = f(-x)$.

This result shows first that the sine function is an odd function ▷; second, that the angle $-C$ corresponds to a clockwise rotation of the point P through an angle C.

The cosine addition formula

From triangle PQS in Figure A2.1, it follows that

$$\cos C = \frac{a}{d} = \frac{r}{r} \cdot \frac{a}{d} = \frac{a^2 + ab}{rd}, \tag{A2.11}$$

where $a + b = r$. But, from the application of the Pythagoras theorem to triangles PQS and SQR:

$$a^2 + h^2 = d^2 \text{ and } b^2 + h^2 = c^2; \tag{A2.12}$$

thus substituting for a^2 and then h^2 from Equation (A2.12) into Equation (A2.5), and using the identity $b^2 + ab = (a + b)^2 - a^2 - ab$ yields

$$\cos C = \frac{d^2 - c^2 + b^2 + ab}{rd} = \frac{d^2 - c^2 + r^2 - a^2 - ab}{rd}$$
$$= \frac{d^2 - c^2 + r^2 - rd\cos C}{rd}.$$

The last step involves using Equation (A2.11) again in the form $rd\cos C = a^2 + ab$. Hence, on collecting the two terms in $\cos C$:

$$\cos C = \frac{d^2 + r^2 - c^2}{2rd}. \tag{A2.13}$$

As seen in Chapter 15, another representation of this result is available from the properties of vectors since if

$$d = |\overrightarrow{PQ}| = |\mathbf{d}|$$
$$c = |\overrightarrow{QR}| = |\mathbf{c}|$$
$$r = |\overrightarrow{PR}| = |\mathbf{r}|,$$

then, from the triangle rule (Figure 15.3), $\mathbf{c} = |\mathbf{r} - \mathbf{d}|$ and

$$\mathbf{c} \cdot \mathbf{c} = \mathbf{r} \cdot \mathbf{r} + \mathbf{d} \cdot \mathbf{d} - 2\mathbf{r} \cdot \mathbf{d} \Rightarrow c^2 = r^2 + d^2 - 2rd\cos C. \tag{A2.14}$$

Thus, for the given vectors $\mathbf{r}$ and $\mathbf{d}$, the distance between their end-points is just given by $c = |\mathbf{r} - \mathbf{d}|$.

Equation (A2.13) can be applied to any of the interior angles in a triangle: the rule involves first summing the squares of the lengths of the two sides defining a given angle, then subtracting the square of the length of the side opposite to the angle before dividing by the product of the lengths of the two sides defining C. It follows, therefore, that application of the rule to the angle $A + B$ in triangle PQR (Figure A2.1), using the identities given in Equation (A2.12), yields

$$\cos (A + B) = \frac{d^2 + c^2 - r^2}{2cd} = \frac{a^2 + b^2 + 2h^2 - r^2}{2cd}$$

$$= \frac{2h^2 - 2ab}{2cd} \text{ (substituting } r = a + b) = \frac{h^2 - ab}{cd}$$

$$= \frac{h}{c} \cdot \frac{h}{d} - \frac{a}{d} \cdot \frac{b}{c} \tag{A2.15}$$

$$= \cos A \cos B - \sin A \sin B, \tag{A2.16}$$

where, in proceeding from Equation (A2.15) to (A2.16), use is made of the definitions (Figure A2.1)

$$\sin A = a/d, \ \sin B = b/c, \ \cos A = h/d, \ \cos B = h/c. \tag{A2.17}$$

More properties of the cosine function

If $A = 2\pi$ and $B = -C$ are substituted in Equation (A2.16) then the equality $\cos (2\pi - C) = \cos (-C)$ is obtained. Furthermore, as the values of x/r associated with $(2\pi - C)$ and C are the same, it follows that

$$\cos (-C) = \cos C, \tag{A2.18}$$

thereby indicating that the cosine function is even in character.

If A is substituted for B in Equation (A2.16), then

$$\cos 2A = \cos^2 A - \sin^2 A. \tag{A2.19}$$

Finally, since $a^2 + h^2 = d^2$,

$$\frac{a^2}{d^2} + \frac{h^2}{d^2} = 1 \Rightarrow \left[\frac{a}{d}\right]^2 + \left[\frac{h}{d}\right]^2 = 1 \Rightarrow \cos^2 A + \sin^2 A = 1.$$

Therefore, from Equation (A2.19),

$$\cos 2A = 1 - 2 \sin^2 A \Rightarrow \sin^2 A = \frac{1}{2}(1 - \cos 2A)$$

or

$$\cos 2A = 2 \cos^2 A - 1 \Rightarrow \cos^2 A = \frac{1}{2}(1 + \cos 2A).$$

Different forms of the addition formulae

The addition formulae for sine and cosine are derived above for the sum of two angles; the corresponding formula for the difference of two angles follows directly from Equations (A2.5) and (A2.16), and the even and odd characters of the cosine and sine functions, respectively:

$$\sin(C - D) = \sin(C)\cos(-D) + \cos(C)\sin(-D)$$
$$= \sin C \cos D - \cos C \sin D$$
$$\cos(C - D) = \cos(C)\cos(-D) - \sin(C)\sin(-D)$$
$$= \cos C \cos D + \sin C \sin D.$$

Inverse trigonometrical functions revisited

Using triangle PQS, it follows that $a^2 + h^2 = d^2$ and

$$\sin C = h/d, \ \cos C = a/d, \ \tan C = h/a. \qquad (A2.20)$$

Thus,

$$C = \arcsin(h/d) = \arccos(a/d) = \arctan(h/a). \qquad (A2.21)$$

Analogous expressions using, for example, $d = \sqrt{a^2 + h^2}$, are entirely equivalent:

$$C = \arcsin\left(\frac{h}{\sqrt{a^2 + h^2}}\right) = \arccos\left(\frac{a}{\sqrt{a^2 + h^2}}\right). \qquad (A2.22)$$

It also follows from Equation (A2.21) (see also Figure A2.2) that $\sin C$ and $\cos C$ may be written in the forms

$$\sin C = \sin(\arccos(a/d)) = \sin(\arctan(h/a)) = \frac{h}{\sqrt{a^2 + h^2}} ; \qquad (A2.23)$$

$$\cos C = \cos(\arcsin(h/d)) = \cos(\arctan(h/a)) = \frac{a}{\sqrt{a^2 + h^2}} . \qquad (A2.24)$$

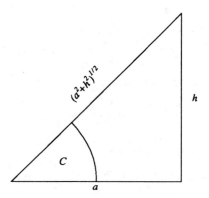

Figure A2.2
A right-angled triangle with base a and height h

It should be noted that in determining an antiderivative function requiring the use of several substitutions (see Chapter 6), the answer may sometimes appear in any of the above convoluted-looking forms; in these situations further simplications can be made using the identities of the kind given in Equations (A2.23) and (A2.24).

Appendix 3

Derivatives of selected functions

$f(x)$	$f'(x)$
x^n	nx^{n-1}
e^{ax}	ae^{ax}
$\ln ax$	$1/x$
$\sin ax$	$a \cos ax$
$\cos ax$	$-a \sin ax$
$\tan ax$	$a \sec^2 ax$
$\cot ax$	$a \operatorname{cosec}^2 ax$
$\sinh ax$	$a \cosh ax$
$\cosh ax$	$a \sinh ax$
$\tanh ax$	$a \operatorname{sech}^2 ax$
$\arcsin ax$ or $\sin^{-1} ax$	$a/(1 - a^2x^2)^{1/2}$
$\arccos ax$ or $\cos^{-1} ax$	$-a/(1 - a^2x^2)^{1/2}$
$\arctan ax$ or $\tan^{-1} ax$	$a/(1 + a^2x^2)$

Note that indefinite integrals of the functions under the column headed $f'(x)$ are given by adding a constant to the entry under the column headed $f(x)$; the indefinite integral of $1/x$ is just $\ln x + C$, since $\ln ax = \ln a + \ln x$, and $\ln a$ can be absorbed in the constant of integration.

Answers to problems

▷
Smart and Moore,
p. 101.

1 Numbers, symbols and rules

1.1 11/2000, 1/18, 1/27;

1.2 (a) 1, 3/2, 3/4, (b) 21, 5/8, 25/6;

1.3 5/9, 6/5, 15/8, 3;

1.4 (a) 2/3, $x^{\frac{4}{3}}$, 1/125, $3^{\frac{8}{3}}$, ab, $2m^{\frac{1}{2}}x^2y^{\frac{3}{2}}$, $4xy$, (b) $R = Nk$, $M = Nm$ where N is the Avogadro constant, and $M = 0.028$ kg mol^{-1}, $\bar{c} = 474.7$ m s^{-1};

1.5 (a) 6.888, 2×10^{-6}, 6.12×10^{-3}, (b) 4.6179×10^{-3} K^{-1}
(c) 14 kJ kg^{-1} K^{-1}, (d) 140 pm, (e) 176 000 m^{-1};

1.6 (a) 1700m^2, (b) -2.18×10^{-18} J $= -2.18$ aJ;

1.7 (a) $(\sqrt{5} - 1)/4$, (b) $(1 - \sqrt{5})/2$, (c) $(1 + \sqrt{x})$;

1.8 (a) $x^2/(x + 1)$, (b) $2x^2/(x^2 - 1)$;

1.9 $1x^0 + 4x^1 + 6x^2 + 4x^3 + 1x^4$;

1.10 (a) k, (b), (c) $\binom{N}{k}$, (d) $\binom{N}{k}^2$ ▷;

1.11 $x = 1/2$, $x = (3 \pm \sqrt{5})/2$, $x = 4$, or ± 1, no solutions in the set of real numbers;

1.12 $r = 3(3 \pm \sqrt{3})a_0/2 \simeq 1.9a_0$ or $7.1a_0$;

1.13 (b) $w = \pm(1 - \sqrt{5})/2$, $w = \pm(1 + \sqrt{5})/2$,
(c) $\epsilon = \alpha + (1 + \sqrt{5})\beta/2, \alpha + (\sqrt{5} - 1)\beta/2$;

1.14 (a) $x = -1 \pm 3i$, $x^* = -1 \mp 3i$, with $xx^* = 10$, (b) $x = 5 + 5i$, $x^2 = 50i$, $x^3 = -250 + 250i$.

2 Functions of a single variable

2.1 (a) $\mathbb{S} = \{x : x \in (-\infty, -2]\} \cup \{x : x \in [2, \infty)\}$
(or $\mathbb{S} = \mathbb{R} - \{x : x \in (-2, 2))$
$\mathbb{S} = \{x : x \in (-\infty, 0)\} \cup \{x : x \in (0, \infty)\}$ (or $\mathbb{R} - \{0\})$,

(b) $f(4) = 4 + 2\sqrt{3}$, $g(\pm\infty) = 1$;

2.2 $\mathbb{S} = \{x : x \in [-2, 1]\}, f(0) = \sqrt{2}, f(\frac{1}{2}) = \sqrt{5}$ and $f(-1) = 2^{-\frac{1}{2}}$;

2.3 (a) 4, 0, $-8/9$, (b) $1/(1+x)(x \neq 0), (2-t)/(3-t)$,
(d) $\sqrt{15}/4$, $\sqrt{15}/4$, 1; domains: (a) $x \in \mathbb{R}, (b) x \in \mathbb{R} - \{-1\}$,
(c) $x \in \mathbb{R} - \{0\}$, (d) $-1 \leq x \leq 1$,
codomains: (a) $[-\frac{9}{4}, \infty)$,
(b) $(-\infty, \infty)$, (c) $(-\infty, \infty)$, (d) $[0, 1]$,
asymptotic values: (a) ∞, (b) 1, (c) 0;

2.4 (c) $E(0) = 0$, $E(1) = 4.206 \times 10^{-22}$ J, $E(2) = 1.262 \times 10^{-21}$ J,
$E(3) = 2.524 \times 10^{-21}$ J, $E(4) = 4.206 \times 10^{-21}$ J,
(d) $J = 0, 1, 2, 3, 4, \ldots$ (e) see Figure 1, where $E(J)$ is measured in
units of 1.4×10^{-21} J;

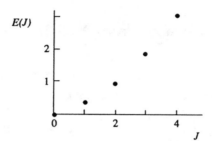

Figure 1
The rotational energy
function for HCl

2.5

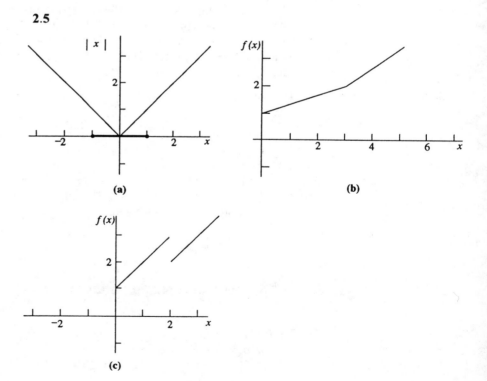

2.6 See Salas and Hille, Section 7.4;

2.7 See Atkins, Figure 19.7; as $T \to \infty$, $n_2/(n_1 + n_2) \to 1/2$;

2.8 (a) take logarithms to the base e and use the result $\ln e^{n \ln x} = n \ln x$,

(b) $\ln(x/y) = \ln e^{p-q} = p - q = \ln x - \ln y$;

2.9 (a) $g(x) = x \ln 3$, $f(x+h) - f(x) = e^{(x+h) \ln 3} - e^{x \ln 3}$
$= (e^{\ln 3})^{x+h} - (e^{\ln 3})^{x} = 3^x(3^h - 1)$, (b) $f(x) = x^2 \ln x$,

(d) $W = 2^N \Rightarrow S = Nk \ln 2 = R \ln 2$;

▷

Barrow, Section 18-6.

2.10 (a) $\ln k = \ln A - E/RT$, (d) $\ln A = 26.22$ or $A = 2.44 \times 10^{11}$ mol^{-1} dm^3 s^{-1} ▷ (a rough error estimate is obtained by drawing lines of maximum and minimum slopes);

2.11 (a) $\cos 2A = \cos(A + A) = \cos^2 A - \sin^2 A = 2 \cos^2 A - 1$,
$\sin 2A = \sin(A + A) = 2 \sin A \cos A$,

(b) $\cos 3A = \cos(2A + A) = \cos 2A \cos A - \sin 2A \sin A =$
$2 \cos^3 A - \cos A - 2(1 - \cos^2 A) \cos A = 4 \cos^3 A - 3 \cos A$,
$\sin 3A = \sin(2A + A) = \sin 2A \cos A + \cos 2A \sin A$
$= 2 \sin A(1 - \sin^2 A) + (1 - 2 \sin^2 A) \sin A = 3 \sin A - 4 \sin^3 A$,

(c) $1 + \tan^2 A = 1 + \dfrac{\sin^2 A}{\cos^2 A} = \dfrac{\cos^2 A + \sin^2 A}{\cos^2 A} = \dfrac{1}{\cos^2 A} = \sec^2 A$,

$1 + \cot^2 A = 1 + \dfrac{\cos^2 A}{\sin^2 A} = \dfrac{1}{\sin^2 A} = \text{cosec}^2 A$,

(d) $\tan (A + B) = \dfrac{\sin (A + B)}{\cos (A + B)}$; expand numerator and denominator,
then divide numerator and denominator by $\cos A \cos B$ to give the result;

2.12 (a) $\frac{1}{2}(e^x - e^{-x}) + \frac{1}{2}(e^x + e^{-x}) = e^x$, $\frac{1}{2}(e^x + e^{-x}) - \frac{1}{2}(e^x - e^{-x}) = e^{-x}$,
$\ln(\cosh x + \sinh x) = \ln e^x = x = -\ln e^{-x} = -\ln(\cosh x - \sinh x)$,
(b) $\cosh^2 x - \sinh^2 x = (\cosh x + \sinh x)(\cosh x - \sinh x) =$
$e^x \cdot e^{-x} = 1$,
$\cosh^2 x + \sinh^2 x = \frac{1}{2}\left((\sinh x + \cosh x)^2 + (\cosh x - \sinh x)^2\right) =$
$\frac{1}{2}(e^{2x} + e^{-2x}) = \cosh 2x$, $\frac{1}{2}(e^{2x} - e^{-2x}) = \frac{1}{2}(e^x + e^{-x})(e^x - e^{-x}) =$
$2 \sinh x \cosh x$;

2.13 (a) $\dfrac{\sinh^2 x}{\cosh^2 x} + \dfrac{1}{\cosh^2 x} = \dfrac{\cosh^2 x}{\cosh^2 x}$ (from Problem 2.12(b)) $= 1$,

$\dfrac{\cosh^2 x}{\sinh^2 x} - \dfrac{1}{\sinh^2 x} = \dfrac{\cosh^2 x - 1}{\sinh^2 x} = 1$ (from Problem 2.12(b));

2.14 (a) use $N_1 + N_2 = N$ to eliminate first N_2 and then N_1, (b) $x = N_1 - N_2$,
(c) substitute $x = \Delta/(2kT)$ in the identity given in Problem 2.13(b);

2.15 (b) $w^2 - 2yw - 1 = 0$, (c) determine w by solving the quadratic equation in (b), and then take logarithms to the base e, (d) if $\sqrt{y^2 + 1} \le y$ then the logarithm in (c) is not defined within the real number system,
(e) $y = \ln (x + \sqrt{x^2 + 1})$, (f) the domain for y follows from the requirement $(1 + x)/(1 - x) > 0$.

3   Limits, small steps and smoothness

3.1 (a) Converges, since $\dfrac{3 + \sqrt{1 - 1/n^2}}{2 + 5/n} \to 3/2$, as n increases without limit, (b) does not converge since $\sin n$ oscillates between $+1$ and -1 without reaching any limit, (c) converges, since $(\cos n)/(n + 1)$ can never be greater than $1/(n + 1)$ nor less than $-1/(n + 1)$, both values of which tend to zero for large n; the situation is unchanged if the domain is extended to include negative integers;

3.2 (a) $1/2$, two discontinuities, both infinite, at $x = \pm\sqrt{2}$; 7, continuous everywhere; -2, one discontinuity, infinite, at $x = 3$, (b) 4, function not defined at $x = 2$; 0, function not defined at $x = 2$; $1/2$, one discontinuity, infinite, at $x = \frac{1}{2}$; 1, continuous all x;

3.3 (a) first order, (b) zeroth order;

3.4 (a) $[\mathrm{Br_2}]^{\frac{1}{2}}$, (b) $[\mathrm{Br_2}]^{\frac{3}{2}}$;

3.5 (a) $3, 3, f(2) = 3$ – results which show that $f(x)$ is continuous at $x = 2$, (b) see Figure 2;

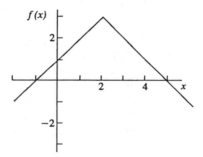

Figure 2

3.6 (b) at high T, use the expansion for e^x given in Example 3.3, (c) $b \to \alpha k_1 t e^{-k_1 t}$.

4   Differentiation and rates of change

4.1 (a) 0, since $f(x) - f(x + h) = 0$, (c) from the hint $[(x + h)^n - x^n]/h$ may be written as $nx^{n-1} + n(n - 1)x^{n-2}h/2 + \cdots$, yielding nx^{n-1} as the derivative. The derivatives for $n = 2, -3, 5/2$ are $2x$, $-3x^{-4}$ and $5x^{3/2}/2$, respectively;

4.4 (a) $1 + \cos x$, (b) $9x^2 - 18x + 54$, (c) $2e^{2x}$, (d) $(x^2 - 2x)/(x - 1)^2$, (e) $-\sin 2x$, (f) $\ln x + 1$, (g) $\sec^2 x$, (h) $e^{2x}((2x + 1) \ln x + 1)$, (i) $2 \tan x \sec^2 x$, (j) $\cosh x$, (k) $\sinh x$, (l) $2/(1 + \sin 2x)$;

4.5 (a) $\cos x \, e^{\sin x}$, $h'(x)e^{h(x)}$, $2 \cos (x/2)$, (b) $1/(x \ln x)$, $12(1 + 3x)^3$, $-6x \sin (3x^2 + 1)$;

4.6 $h'(J) = \left[(2 - (2J + 1)^2 \Theta_r/T\right] e^{-J(J+1)\Theta_r/T}/q_r$;

4.7 (a) $f^{(n)}(x) = (-1)^{(n-1)}(n-1)!\,(1+x)^{-n}$, not defined for $x = -1$,

(b) $f^{(n)}(x) = a^n e^{ax}$, (c) $f^{(n)}(x) = \cos(x + (n-1)\pi/2)$, $n = 1, 2, \ldots$,

(d) $f^{(n)}(x) = (-1)^{n-1} \dfrac{(2n-3)!}{2^{2n-2}(n-2)!}(1+x)^{-\frac{2n-1}{2}}$, $n = 2, 3, \ldots$;

4.8 $\dfrac{d\phi_i}{dr} = -\dfrac{Z_i}{r} e^{-r/r_D}\left(\dfrac{1}{r} + \dfrac{1}{r_D}\right)$;

4.9 (a) Maximum: $r = 11.4686\,a_0/Z, \psi = 1.828\,41N$; minimum: $r = 3.5314\,a_0/Z, \psi = -3.581\,95N$, (b) see Alberty and Silbey, Section 11.2 ;

4.10 $J = 3$ is the nearest integer ;

4.11 (b) The second derivative of V has the positive value $1/\rho_m^3$ at ρ_m, indicating that the turning point is a minimum ;

4.12 (a) $r_m = 2^{\frac{1}{6}}\sigma$;

4.13 (a) The turning points occur when $\cos(n\pi x/L) = 0$,

(b) $x = L/6, L/2$, and $5L/6$, (c) the turning points in part (b) are maximum, minimum and maximum, respectively ;

4.14 (a) $b'(t) = \dfrac{\alpha \cdot k_1}{k_2 - k_1}(-k_1 e^{-k_1 t} + k_2 e^{-k_2 t})$,

(c) $b'' = \dfrac{\alpha \cdot k_1}{k_2 - k_1}(k_1^2 e^{-k_1 t} - k_2^2 e^{-k_2 t})$ and at $t = t_m$ has the value $-\alpha k_1^2 e^{k_1 t_m}$ which is always negative ;

4.15 (a) $\left(\dfrac{\partial P}{\partial T}\right)_V = R/V$ and $\left(\dfrac{\partial P}{\partial V}\right)_T = -RT/V^2$, (b) $\left(\dfrac{\partial V}{\partial T}\right)_P = R/P$,

(c) $\alpha = \dfrac{1}{V}\left(\dfrac{\partial V}{\partial T}\right)_P = 1/T$;

4.16 (a) $T \geq 0$ K, and $\dfrac{V^2}{V-b} \geq \dfrac{a}{RT}$ (the latter relation subsumes the requirement $V > b$), (b) $\left(\dfrac{\partial P}{\partial V}\right)_T = -\dfrac{RT}{(V-b)^2} + \dfrac{2a}{V^3}$,

$\left(\dfrac{\partial P}{\partial T}\right)_V = \dfrac{R}{V-b}$, $\left(\dfrac{\partial^2 P}{\partial V^2}\right)_T = \dfrac{2RT}{(V-b)^3} \dfrac{6a}{V^4}$

---

5   Differentials – small and not so small changes

5.1 $dy = 2.0, \Delta y = 2.25$, $dy = 0.2, \Delta y = 0.2025$,

$dy = 8.0, \Delta y = 9.875$, $dy = 0.8, \Delta y = 0.8176$;

5.2 $dy = 0.555\ldots, \Delta y = 0.5108$, $dy = 0.0055\ldots, \Delta y = 0.005\,550$;

5.3 (a) $dC_p = C_p' dT = (b + 2cT)dT$,

(b) $dC_p = 0.03046$ J K^{-1} mol^{-1}, $\Delta C_p = 0.030\,62$ J K^{-1} mol^{-1} ;

5.4 (a) λ at 298 K is 2.5237×10^{-10} m and at 288 K is 2.5671×10^{-10} m,

(b) $d\lambda = -\dfrac{h}{2(mkT^3)^{\frac{1}{2}}}\,dT$,

(c) $d\lambda = 4.23 \times 10^{-12}$ m, $\Delta\lambda = 4.34 \times 10^{-12}$ m ;

5.5 (a) Use the identity $10^x = e^{x \ln 10}$, (b) $K = 0.472$ for $I/\text{mol kg}^{-1} = 0.025$
and $K = 0.558$ for $I/\text{mol kg}^{-1} = 0.015$;

5.6 (a) 1.895 49, (b) all the roots are complex, (c) 2.828,
(d)(i) $\rho = 1.268$, (ii) 4.732, (iii) 4.732;

5.7 $e^x \simeq 1 + h = 1.01$, exact is 1.010 0502, percentage error $< 5 \times 10^{-3}$;

5.8 (b) $\alpha = \dfrac{W'(T_0)}{W(T_0)}$, (d) $dV = 1.05 \times 10^{-4}\ \text{dm}^3$, $100 \times dV/V = 0.105\%$;

5.9 (a), (b) use $a = \alpha e^{-kt} \Rightarrow \alpha/2 = \alpha e^{-k\tau} \Rightarrow e^{k\tau} = 2$;

5.10 (a) $dV = \left(\dfrac{\partial V}{\partial n_A}\right)_{n_B,T,P} dn_A + \left(\dfrac{\partial V}{\partial n_B}\right)_{n_A,T,P} dn_B + \left(\dfrac{\partial V}{\partial T}\right)_{n_A,n_B,P} dT$

$+ \left(\dfrac{\partial V}{\partial P}\right)_{n_A,n_B,T} dP,$

(b) T, P constant: $dV = \left(\dfrac{\partial V}{\partial n_A}\right)_{n_B} dn_A + \left(\dfrac{\partial V}{\partial n_B}\right)_{n_A} dn_B,$

(c) T, P, n_A constant: $dV = \left(\dfrac{dV}{dn_B}\right) dn_B.$

6   Integration – undoing the effects of differentiation

6.1 (a) $\frac{1}{2}e^{x^2} + C$, (b) $\frac{1}{2}\ln(\ln x) + C$, (c) $2x^{1/2} + C$;

6.2 (a) $\frac{1}{12}\sin 3x + \frac{3}{4}\sin x + C = \sin x - \frac{1}{3}\sin^3 x + C$,
(b) $\frac{1}{4}\left(\frac{3}{2}x + \sin 2x + \frac{1}{8}\sin 4x\right) + C$;

6.3 (a) $-1 < x < 1$, $\sqrt{1-x^2} + C$ (or use $x = \cos\theta$ or $e^u = 1 - x^2$),
(b) $x \geq -4$, $\frac{2}{5}(x+4)^5 - \frac{8}{3}(x+4)^3 + C$ (or use $x = 4\sinh^2\theta$ or
$x = 4\tan^2\theta$),   (c) $x > 0$, $x \neq 1$, $2\ln(x^{\frac{1}{2}} - 1)$ (or use $x = \sin^4\theta$),
(d) $-\infty < x < \infty$, $\frac{1}{a}\tan^{-1}(x/a) + C$ (or $x = a\sinh u$ and the result
in (f)),

(e) $-\infty < x < \infty$, $[x/(x^2 + a^2)^{\frac{1}{2}} - \frac{1}{3}x^3/(x^2 + a^2)^{\frac{3}{2}}]/a^4 + C$,
(f) $-\infty < x < \infty$, $x \neq 0$, $2\tan^{-1}(e^x) + C$,
(g) $-\infty < x < \infty$, $\ln(e^x - 1) + C$ (or $e^x = \cos^2\theta$), (h) $-\infty < x < \infty$,
$\frac{\sin^4 x}{4} - \frac{\sin^6 x}{6} + C$ (or $u = \cos x$);

6.4 $f(x) = x^2 + 4$, $f'(x) = 2x$, $\int \frac{x}{x^2+4}\,dx = \frac{1}{2}\ln(x^2 + 4) + C$;
$f(x) = x + 1$, $f'(x) = 1$, $\int \frac{1}{x+1}\,dx = \ln(x + 1) + C$;
$f(x) = \cos x$, $f'(x) = -\sin x$, $\int \tan x\,dx = -\ln(|\cos x|) + C$,
$f(x) = \ln x$, $f'(x) = 1/x$, $\int \frac{1}{x \ln x}\,dx = \ln(\ln x) + C$;

6.5 (a) $\frac{1}{(b-a)}\ln[(b - x)/(a - x)] + C$, remembering that
$\int \frac{1}{(a-x)}\,dx = -\ln(a - x)$, using $u = a - x$;
(b) $\frac{1}{3}(\ln(x + 1)) + \frac{2}{3}(\ln(x - 2)) + C$;

6.6 (a) $A + D = 0$, $2A + B + C + D = 0$, $A + 2B + 2D = 1$,
$B + 2C + 2D = 0$,

(b) $\frac{1}{9}\{-\frac{1}{2}\ln(x^2+2)+2\sqrt{2}(\arctan(x/\sqrt{2}))+3/(x+1)$
$+\ln(x+1)\}+C$;

6.7 (c) See the integral following Equation (6.7);

6.8 $f(x)=x,\ g'(x)=e^{-x},\ \int xe^{-x}\,dx=-e^{-x}(x+1)+C$;
$f(x)=x,\ g'(x)=xe^{-x},\ \text{or}\ f(x)=x^2,g'(x)=e^{-x},$
$\int x^2e^{-x}dx=-e^{-x}((x+1)^2+1)+C$;
$f(x)=x,\ g'(x)=\sin x,\ \int x\sin x\,dx=\sin x-x\cos x+C$;
$f(x)=\ln x,\ g'(x)=x^2,\ \int x^2\ln x\,dx=\frac{x^3}{3}\ln x-\frac{x^3}{9}+C$;
$f(x)=\cos x,\ g'(x)=\sin x,\ I=\int\sin x\cos x\,dx=$
$-\cos^2 x-I+C'\Rightarrow I=-\frac{1}{2}\cos^2 x+C$;
$f(x)=1,\ g'(x)=\ln x,\ \int\ln x\,dx=x\ln x-x+C$;

6.9 $\int_0^{\pi/2}\cos x\,dx=[\sin x]_0^{\pi/2}=1$;

6.10 $I=\int_a^b f(x)dx=F(b)-F(a)$; $I_1=\int_a^c f(x)dx=F(c)-F(a)$;
$I_2=\int_c^b f(x)dx=F(b)-F(c)$; thus, $I=I_1+I_2$;

6.11 (a) $5/14,\ 20,\ \frac{1}{4}\ln(25/9)\ \triangleright,\ \ln(81/5)$,
(c) $(1-e^{-2L})/2,\ -e^{-2L}(L^2/2+L/2+1/4)+1/4\ \triangleright$,
(d) $a(T_2-T_1)+b(T_2^2-T_1^2)/2+c(T_2^3-T_1^3)/3$;

6.12 (a) $x=0$ and $x=1,\ f(0)=0,\ f(1)=4/e^2$, (b)
$e^{-2x}(16x^2-32x+8)$; $x=0$, a minimum, $x=1$, a maximum; (c) see
Figure 3;
(e) $f(1)=4/e^2$ and therefore
$4/2e^2<1-5/e^2<4/e^2\Rightarrow 7<e^2<9\Rightarrow\sqrt{7}<e<3$;

$\triangleright$
Use partial fractions.
$\triangleright$
Substitute $x=2r$ and
use the result for the
second integral in
Problem 6.8.

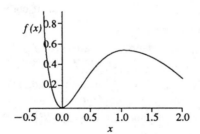

Figure 3
Plot of the function
$f(x)=4x^2e^{-2x}$ for the
interval $-0.5<x<2.0$

6.14 (a) $\ln K(T)=-\Delta H^\ominus/(RT)+C$,
(b) $\ln K(T)=(c(T_2-T_1)+b\ln(T_2/T_1)-a(1/T_2-T_1))/R$;

6.16 (a) Use partial fractions, and then break the range of integration at $x=2$
for the integral involving the integrand $1/(x-2)$:
$I=\frac{1}{4}\lim_{\epsilon\to 0}\{[\ln|x-2|]_0^{2-\epsilon}+[\ln|x-2|]_{2+\epsilon}^7\}-(1/4)\{\ln 9-\ln 2\}=$
$\ln(5/9)$, (b) $1/4$, taking the limit $L\to\infty$ in the answer to the second
integral in Problem 6.11(c), (c) $4/(3a^4)$;

$\triangleright$
This substitution does
not change the values of
the limits.

6.17 (a) $\langle v\rangle=2\pi(m/(2\pi kT))^{3/2}\int_0^\infty e^{-\alpha u}u\,du\ \triangleright$;

6.18 (a) T/Θ_r;

6.19 Using $A=13.01/23^3$ J mol^{-1} K^{-4} to ensure continuity at $T=23$ K,
$S=127.6$ J mol^{-1} (found using five trapezia and one triangle to
approximate the area under the curve).

7   Power series: a new look at functions

7.1 (a) $xS_r = x + x^2 + \cdots + x^r + x^{r+1}$, $S_r - xS_r = 1 - x^{r+1}$ and hence result; (b) if $|x| < 1$, then $\lim_{r \to \infty} x^r = 0$, (c) 2, $1/(1 - e^\theta)$;

7.2 (a) Replace x by $-w$, (b) q_v is the form of Equation (7.1) with $x = e^{-hv/kT}$; use the result in Problem 7.1(b),
(c) $(3 - 2x)^{-1} = \frac{1}{3}(1 - \frac{2}{3}x)^{-1} = \sum_{m=0}^{\infty}(2^m/3^{m+1})x^m$;

7.3 Use Example 7.2 with $a = 2$ then $f^{(n)}(2) = (-1)^n n!/4^{(n+1)}$, and the ratio test yields $\lim_{r \to \infty}|(x - 2)/4)| < 1$ for convergence, i.e., $-2 < x < 6$, $f(0.9) = 0.3448$ for the seventh-degree polynomial;

7.4 $5 + 6(x - 1) + 3(x - 1)^2 + (x - 1)^3$;

7.5 $e^x = 1 + x + x^2/2! + \cdots + x^r/r! + \cdots$, converges $-\infty < x < \infty$;
$e^{-x} = 1 - x + x^2/2! + \cdots + (-1)^r x^r/r! + \cdots$, converges $-\infty < x < \infty$;
$\cos x = 1 - x^2/2! + x^4/4! + \cdots + (-1)^r x^{2r}/(2r)! + \cdots$, converges $-\infty < x < \infty$; $\sin x = x - x^3/3! + \cdots + (-1)^r x^{2r+1}/(2r + 1)! + \cdots$, converges $-\infty < x < \infty$;
$\ln(1 + x) = x - x^2/2 + x^3/3 + \cdots + (-1)^r x^{r+1}/(r + 1) + \cdots$, converges $-1 < x < 1$ ▷;

▷
In all cases $r = 0$ designates the first term.

7.6 2, 4;

7.7 (b) $\sum_{r=1}^{\infty}(-1)^{r+1}x^{2r-1}/(2r - 1)!$, (c) $\sum_{r=0}^{\infty}(-1)^r x^n$, $(1 + x)^{-1}$, with interval of convergence $-1 < x < 1$,
(d) $\sum_{r=1}^{\infty}(-1)^{r-1}\frac{x^{r+1}}{r(r+1)} + C$, interval of convergence $-1 < x < 1$,
(e) $(x + 1)\ln(x + 1) - x + C$;

▷
0! is defined as 1

7.9 (a) $\sum_{r=1}^{\infty}\frac{x^{r+1}}{(r-1)}$, converges for all x, (b) $1 + 2\sum_{r=1}^{\infty}x^r$, converges for $-1 < x < 1$, (c) $(\sum_{m=0}^{\infty}x^m)(\sum_{p=0}^{\infty}x^p/p!) = \sum_{r=0}^{\infty}a_r x^r$, where ▷
$a_r = \frac{1}{0!} + \frac{1}{1!} + \frac{1}{2!} + \cdots + \frac{1}{r!}$ (after collecting all terms with the same power of x), with interval of convergence $-1 < x < 1$;

▷
Since this test involves evaluating the ratio of successive terms in the series for increasingly large r, the test here is applied to the $(r + 2)$th and $(r + 1)$th terms.

7.10 (b) The ratio test yields ▷ $\lim_{r \to \infty}\left|\frac{u_{r+2}}{u_{r+1}}\right| =$
$\lim_{r \to \infty}\left|\frac{\alpha(\alpha-1)(\alpha-2)\cdots(\alpha-r+1)}{r!} \cdot \frac{(r-1)!}{\alpha(\alpha-1)(\alpha-2)\cdots(\alpha-r+2)} \cdot x\right| = \lim_{r \to \infty}\left|\frac{(\alpha-r+1)x}{r}\right| =$
$|x|$, which yields the interval of convergence $-1 < x < 1$,
(c) $\alpha = -1 : \sum_{n=0}^{\infty}(-1)^n x^n$; $\alpha = \frac{1}{2}$:
$1 + \frac{x}{2} + \sum_{n=2}^{\infty}\frac{(-1)^{n-1}(2n-3)(2n-5)\cdots 5 \cdot 3 \cdot 1}{2^n n!}x^n$, (d) 0.990, 1.049;

7.11 (a) $E(R) = E(R_e) + E^{(1)}(R_e)(R - R_e) + E^{(2)}(R_e)(R - R_e)^2/2! + \cdots$
$+ E^{(n)}(R_e)(r - R_e)^n/n! + \cdots$,
(b) $E^{(1)}(R_e)$, (c) degree 2: $E(R) = E(R_e) + E^{(2)}(R_e)(R - R_e)^2/2$,
degree 3: $E(R) = E(R_e) + E^{(2)}(R_e)(R - R_e)^2/2 + E^{(3)}(R_e)(R - R_e)^3/3!$
(d) $\mathcal{EE}(R) = E^{(2)}(R_e)(R - R_e)^2/2 = \frac{1}{2}k(R - R_e)^2$, where $k = E^{(2)}(R_e)$;

7.12 (a) $\mathcal{E}^{(1)}(R_e) = E^{(1)}(R_e) = D_e(2\alpha e^{-\alpha(R-R_e)} - 2\alpha e^{-2\alpha(R-R_e)})$,
$\mathcal{E}^{(2)}(R_e) = E^{(2)}(R_e) = -D_e(-2\alpha^2 e^{-\alpha(R-R_e)} + (-2\alpha)^2 e^{-2\alpha(R-R_e)})$,
$\mathcal{E}^{(3)}(R_e) = E^{(3)}(R_e) = D_e(2\alpha^3 e^{-\alpha(R-R_e)} + (-2\alpha)^3 e^{-2\alpha(R-R_e)})$,

$$\mathcal{E}^{(4)}(R_e) = E^{(4)}(R_e) = D_e(2\alpha^4 e^{-\alpha(R-R_e)} + (-2\alpha)^4 e^{-2\alpha(R-R_e)})\,;\text{ thus, for}$$
$r \neq 0,\ \mathcal{E}^{(r)}(R_e) = E^{(r)}(R_e)D_e((-1)^{r+1}2\alpha^r e^{-\alpha(R-R_e)} + (-2\alpha)^r e^{-2\alpha(R-R_e)})$,
and the derivatives at R_e are 0, $2D_e\alpha^2$, $-6D_e\alpha^3$,
$14D_e\alpha^4,\ldots 2D_e(-1)^r\alpha^r(2^{r-1}-1),\ldots$,

(b) $\mathcal{E}(R) = \sum_{r=2}^{\infty} 2D_e(-1)^r(2^{r-1}-1)\alpha^r(R-R_e)^r/r!$,

(d) $a = 2\pi v\sqrt{\mu/(2D_e)}$, (f) see Atkins, Figure 16.37.

8   Complex numbers revisited

8.1 $z_1z_2 = 4 - i\sqrt{2}$, $z_1 + 2z_2 = 5 + i\sqrt{2}$, $z_1 - z_2 = -1 - i2\sqrt{2}$,
$z_3 = (1 + i\sqrt{2})/3$;

8.2 (a) $6 - 6i$, (b) $-i$, (c) $2i$, (d) $(1 - 2i)/5$, (e) $\cos\beta - i\sin\beta$;

8.3 See also Figure 4. Problem 8.1: $4 - i\sqrt{2}, r_1 = \sqrt{18}, \theta_1 = -0.3398^c$;
$5 + i\sqrt{2}, r_2 = \sqrt{27}, \theta_2 = 0.2756^c$; $-1 - i2\sqrt{2}, r_3 = 3, \theta_3 = -1.2310^c$;
$(1 + i\sqrt{2})/3, r_4 = 1/\sqrt{3}, \theta_4 = 0.9553^c$; Problem 8.2:
(a) $6 - 6i, r_5 = 6\sqrt{2}, \theta_5 = -0.7854^c$, (b) $-i, r_6 = 1, \theta_6 = -\pi/2^c$,
(c) $2i, r_7 = 2, \theta_7 = \pi/2^c$, (d) $(1 - 2i)/5, r_8 = 1/\sqrt{5}, \theta_8 = -1.1071^c$,
(e) $\cos\beta - i\sin\beta, r = 1, \theta = -\beta$;

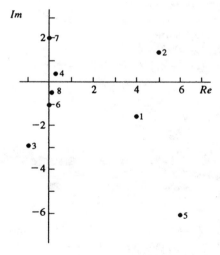

Figure 4
Argand diagram for
Problem 8.3, where r_m and
θ_m are identified with $\bullet_m$,
representing a complex
number $z = x + iy$

8.5 (b) $w^2 - 10w + 1 = 0$ has roots $w = 9.898\,98$ and $w = 0.101\,02$,
(c) $x = \pm 2.2924$;

8.6 (a) $e^{i\pi/2} = -i$, (b) $e^{i2\pi} = 1$, (c) $(1 - i\sqrt{3})^{10}, = 2^9(-1 + i\sqrt{3})$
(for $z = 1 - i\sqrt{3}, r = 2, \theta = -\pi/3$), (d) $(1 - i\sqrt{3})^{-6} = 2^{-6}$;

8.7 (a) $F_1 = \sum_j f_j \cos[2\pi(hx_j + ky_j + lz_j)]$,
$F_2 = \sum_j f_j \sin[2\pi(hx_j + ky_j + lz_j)]$, (b)2 using
$e^{\pi i(2n+1)} = -1, e^{\pi i2n} = 1$, (300) missing, (200) present, (111) missing,

(222) present; (b)3 $f_A \neq f_B$ and hence $F(300)$, $F(111)$, given by $f_A - f_B$, are not zero, (c)1 $F(hkl)$ vanishes if one of hkl is even (odd) and the other pair both odd (even), (c)2 $F(hkl)$ for NaCl vanishes in the same way as for the face-centred structure in (c)1;

8.8 (a) $x = \pm i$, (b) for $i, r = 1, \theta = (4n + 1)\pi/2$, $\sqrt{i} = \pm(1 + i)/\sqrt{2}$, (c) $(1 + i)$, $r = \sqrt{2}$, $\theta = (8n + 1)\pi/4$, $(1 + i)^{\frac{1}{3}} = 2^{\frac{1}{6}}e^{i\pi/12}$, $2^{\frac{1}{6}}e^{3i\pi/4}$, $2^{\frac{1}{6}}e^{-7i\pi/12}$, (d) $2^{\frac{1}{3}}$, $2^{\frac{1}{3}}e^{\pm 2i\pi/5}$, $2^{\frac{1}{3}}e^{\pm 4i\pi/5}$, or 1.1487, $0.3550 \pm 1.0925i$, $-0.9293 \pm 0.6752i$;

8.9 (a) $\ln(-1) = i$, (b) $\ln(1 + i) = \frac{1}{2}\ln 2 + i\pi/4$, (c) $\ln i = i\pi/2$.

9   The solution of simple differential equations – the nuts and bolts of kinetics

9.1 (a) $y = 1/(C - e^x)$, with $C = 2$, (b) $y = A \sin x$, with $A = 4$, (c) $y = \cos^{-1}(x^2/2 + A)$, with $A = 0$, (d) $y = x/(A - x)$, with $A = 2$;

9.2 (a) $a = e^{-kt+C}$ or $a = Ae^{-kt}$, (b) $A = \alpha$;

9.3 (b) $k_{eff} = \beta k$;

9.4 (b) see Atkins, Figure 25.6;

9.5 (d) $y = -(x^2 + 1)/2x$;

9.6 (c) $C = -p_0 - p_0 \ln p_0$;

9.7 (c) $C = -k_1\alpha/(k_2 - k_1)$, $b = k_1\alpha(e^{-k_1 t} - e^{-k_2 t})/(k_2 - k_1)$;

9.8 (a) $y = Ae^{3x} + Be^{2x}$;

9.9 $y = De^{-x} + e^x/2 + C$;

9.10 (h) $A = (2c_0 + c_1)/4$, $B = (2c_0 - c_1)/4$.

10   Functions of two or more variables – differentiation revisited

10.1 (a) $\partial z/\partial x = 6x + 2y^3$, $\partial z/\partial y = 6xy^2 - 6y$, (b) $\partial z/\partial x = 2y/(x + y)^2$, $\partial z/\partial y = -2x/(x + y)^2$, (c) $\partial z/\partial x = -(2/y)\sin(2x/y)$, $\partial z/\partial y = (2x/y^2)\sin(2x/y)$, (d) $\partial z/\partial x = 2xe^{x^2-y^2}(1 + (x^2 + y^2))$, $\partial z/\partial y = 2ye^{x^2-y^2}(1 - (x^2 + y^2))$;

10.2 (a) $C_V = 3R/2$, (b) $C_P = 5R/2$;

10.3 (a) $\partial^2 z/\partial y\partial x = \partial^2 z/\partial x\partial y = 6y^2$, (b) $\partial^2 z/\partial y\partial x = \partial^2 z/\partial x\partial y = 2(x - y)/(x + y)^3$, (c) $\partial^2 z/\partial y\partial x = \partial^2 z/\partial x\partial y = (2/y^2)[\sin(2x/y) + (2x/y)\cos(2x/y)]$, (d) $\partial^2 z/\partial y\partial x = \partial^2 z/\partial x\partial y = -4xye^{x^2-y^2}(x^2 + y^2)$;

10.4 $E = -13.61 \, eV$;

10.5 If $I_n = \displaystyle\int_0^\infty r^n e^{-kr} dr$, then $\partial^2 I/\partial k^2 = I_2 = 2/k^3$,

$\partial^3 I/\partial k^3 = -I_3 = -3 \cdot 2/k^4$, $\partial^4 I/\partial k^4 = I_4 = 4 \cdot 3 \cdot 2/k^5, \ldots$,
$I_n = n \cdot (n-1) \cdot (n-2) \cdots 2/k^{n+1} = n!/k^{n+1}$;

10.6 Partially differentiate twice with respect to k, and then set $k = 1$;

10.7 (b) Critical points arise when $x^2 - y^2 = 0$, and $y = 0$, that is at $(0,0)$,
where $D = -4$ (saddle point), and at $(2,0)$, where $D = 4e^{-4} > 0$
(maximum, since $\partial^2 z/\partial x^2 = -2e^{-2} < 0$);

▷
Note that
$\ln(y/x) = \ln y - \ln x$.

10.8 (a) $dz = (e^y - 1/x)dx + (xe^y + 1/y)dy$ ▷,
(b) $dS = \frac{5}{2}(R/T)dT - (R/P)dP$;

10.9 $da_{H^+} = -\ln 10 e^{-pH \ln 10} dpH \Rightarrow da_{H^+} = \pm 0.01$;

10.11 $dz = (y/z)dy + (x/z)dx = 0.14 \, \text{m}$, using $dx = dy = 0.1 \, \text{m}$;

10.12 (a) $\partial z/\partial v = (6e^{u+v}\cos(e^{u-v}) + 1 + 3e^{(3u+v)}\sin(e^{u-v})$,
(b) $\partial z/\partial s = 10s$, $\partial z/\partial t = -10t$;

10.13 (a) Not exact, (b) exact, (c) exact;

10.14 (a) $F(x,y)$ does not exist, (b) $F(x,y) = -x^2 + y^2/2 + xy + C$,
(c) $F(x,y) = xe^y - y^2 + C$;

10.15 (a) $dP = -(RT/V^2)dV + (R/V)dT$, exact, (b)(i) $W = -P_{ex}(V_2 - V_1)$,
(ii) $-RT \ln(V_2/V_1)$.

---

11 Multiple integrals – integrating functions of several variables

11.2 (a) $\displaystyle\int_0^1 dy \int_0^{\sqrt{1-y^2}} dx \,(xy+1) = \int_0^1 dx \int_0^{\sqrt{1-x^2}} dy \,(xy+1) = 23/24$,
(b) $1/8 + \pi/4$;

11.3 $15/32 + \pi/8$;

11.4 (a) 24, (b) 0, (c) $\displaystyle\int_0^\infty r^2 e^{-r} dr \int_\pi^{2\pi} d\theta = 2\pi$;

11.5 (b) C_V, the heat capacity at constant volume;

11.6 (a) $\int \chi_0(s_1)\mu(R)\chi_1(s_1)\,ds_1 = \mu(R_e)\int \chi_0(s_1)\chi_1(s_1)\,ds_1$

$+\mu^{(1)}(R_e)\int \chi_0(s_1)s_1\chi_1(s_1)\,ds_1 + \cdots$
$= \mu^{(1)}(R_e)\int \chi_0(s_1)s_1\chi_1(s_1)\,ds_1 + \cdots$; the integral is finite, and so the
first derivative of the dipole moment function must be non-zero,

(b) $\int \chi_0(s_1)\mu(R)\chi_0{}'(s_1)\,ds_1 = \mu(R_e)\int \chi_0(s_1)\chi_0{}'(s_1)\,ds_1$

$+\mu^{(1)}(R_e)\int \chi_0(s_1)s_1\chi_0{}'(s_1)\,ds_1 + \cdots$, where the overlap integral
between the vibrational wavefunctions of ground and excited states is
termed a Franck–Condon factor; the intensity is proportional to the
square of this expression.

13   Matrices – a useful tool and a form of mathematical shorthand

13.1 (b) $\mathbf{H} = \begin{pmatrix} 5 & 9 \\ -2 & -2 \\ -4 & 2 \end{pmatrix}$;

13.2 (a) $\mathbf{AD} = \begin{pmatrix} 9 & 20 \\ 4 & 16 \\ -3 & 12 \end{pmatrix}$, (b) $\mathbf{DA}$ not defined, (c) $(\mathbf{CA})\mathbf{D} = (\,25 \quad 100\,)$,

(d) $\mathbf{C}(\mathbf{AD}) = (\,25 \quad 100\,)$, (e) $\mathbf{CE} = (\,7 \quad 18\,)$,

(f) $\mathbf{AE} = \begin{pmatrix} 15 & 14 \\ 5 & 17 \\ 2 & 21 \end{pmatrix}$, (g) $\mathbf{DE}$ not defined, (h) $\mathbf{AC}$ not defined,

(i) $\mathbf{CA} = (\,17 \quad 14 \quad 9\,)$,   (j) $\mathbf{A}(\mathbf{D} + \mathbf{E}) = \begin{pmatrix} 24 & 34 \\ 9 & 33 \\ -1 & 33 \end{pmatrix}$,

(k) $\mathbf{AD} + \mathbf{AE} = \begin{pmatrix} 24 & 34 \\ 9 & 33 \\ -1 & 33 \end{pmatrix}$;

13.3 $(\mathbf{E}_m\mathbf{X})_{ij} = \sum_{k=1}^{m}(\mathbf{E}_m)_{ik}(\mathbf{X})_{kj} = \sum_{k=1}^{m}\delta_{ik}(\mathbf{X})_{kj} = (\mathbf{X})_{ij}$;

13.4 $\mathbf{A}^T = \begin{pmatrix} 4 & 3 & 1 \\ 6 & 0 & -2 \\ -1 & 2 & 5 \end{pmatrix}$, $\mathbf{B}^T = (\,1 \quad 1 \quad 1\,)$,

$\mathbf{C}^T = \begin{pmatrix} 3 \\ 1 \\ 2 \end{pmatrix}$, $\mathbf{D}^T = \begin{pmatrix} 2 & 0 & -1 \\ 4 & 1 & 2 \end{pmatrix}$,

$\mathbf{E}^T = \begin{pmatrix} 1 & 2 & 1 \\ 3 & 1 & 4 \end{pmatrix}$

13.5 (a) $\mathbf{B}^T\mathbf{B}$ is an $m \times n$ times an $n \times m$ (defined), and yields an $m \times m$; $\mathbf{BB}^T$ is an $n \times m$ times an $m \times n$ (defined), and yields an $n \times n$,

(b)
$\mathbf{DD}^T = \begin{pmatrix} 20 & 4 & 6 \\ 4 & 1 & 2 \\ 6 & 2 & 5 \end{pmatrix}$, $\mathbf{D}^T\mathbf{D} = \begin{pmatrix} 5 & 6 \\ 6 & 21 \end{pmatrix}$, $\mathbf{C}^T\mathbf{C} = \begin{pmatrix} 9 & 3 & 6 \\ 3 & 1 & 2 \\ 6 & 2 & 4 \end{pmatrix}$,
$\mathbf{CC}^T = 14$;

13.6 (c) $(\mathbf{C}^T)_{ij} = (\mathbf{C})_{ji} = \sum_{k=1}^{m}(\mathbf{A})_{jk}(\mathbf{B})_{ki} = \sum_{k=1}^{m}(\mathbf{A}^T)_{kj}(\mathbf{B}^T)_{ik} =$

$\sum_{k=1}^{m}(\mathbf{B}^T)_{ik}(\mathbf{A}^T)_{kj} = (\mathbf{B}^T\mathbf{A}^T)_{ij} \Rightarrow \mathbf{C}^T = (\mathbf{AB})^T = \mathbf{B}^T\mathbf{A}^T$;

13.7 (a) $(\mathbf{C})_{pp} = \sum_{k=1}^{m}(\mathbf{A})_{pk}(\mathbf{B})_{kp}$,

(b) $\text{tr}(\mathbf{AB}) = \sum_{p=1}^{m}(\mathbf{C})_{pp} = \sum_{p=1}^{m}(\mathbf{AB})_{pp} = \sum_{p=1}^{m}\sum_{k=1}^{m} a_{pk}b_{kp} =$

$\sum_{k=1}^{m}\sum_{p=1}^{m} b_{kp}a_{pk} = \sum_{k=1}^{m}(\mathbf{BA})_{kk} = \text{tr}(\mathbf{BA})$,

(d) $\text{tr}(\mathbf{ABC}) = \text{tr}(\mathbf{XC}) = \text{tr}(\mathbf{CX}) =$
$\text{tr}(\mathbf{CAB}) = \text{tr}(\mathbf{YB}) = \text{tr}(\mathbf{BY}) = \text{tr}(\mathbf{BCA})$;

13.8 (a) If $\mathbf{C} = \mathbf{AB}$, then

$c_{ij}^* = (\mathbf{AB})_{ij}^* = \left(\sum_{k=1}^{m} a_{ik}b_{kj}\right)^* = \sum_{k=1}^{m} a_{ik}^* b_{kj}^* = (\mathbf{A}^*\mathbf{B}^*)_{ij}$

$\Rightarrow (\mathbf{AB})^* = \mathbf{A}^*\mathbf{B}^*$,

(b) $(\mathbf{C}^{\dagger})_{ij} = c_{ji}^* = \sum_{k=1}^{m} a_{jk}^* b_{ki}^* = \sum_{k=1}^{m}(\mathbf{B}^{\dagger})_{ik}(\mathbf{A}^{\dagger})_{kj} = (\mathbf{B}^{\dagger}\mathbf{A}^{\dagger})_{ij}$;

$\triangleright$
$\mathbf{x}^{\dagger}\mathbf{Ax}$ is a 1×1 matrix.

13.9 (a) $\mathbf{A}^{\dagger} = \mathbf{A}$, (b) If $\mathbf{C} = \mathbf{x}^{\dagger}\mathbf{Ax}$, then
$\mathbf{C}^* = \mathbf{x}^T\mathbf{A}^*\mathbf{x}^* = \mathbf{x}^T\mathbf{A}^T\mathbf{x}^* = (\mathbf{x}^{\dagger}\mathbf{Ax})^T = \mathbf{x}^{\dagger}\mathbf{Ax} = \mathbf{C} \triangleright$,
$\mathbf{x}^{\dagger}\mathbf{Ax} = 2(a_1^2 + b_1^2 + a_2b_1 - a_1b_2) + (a_2^2 + b_2^2) + 6(a_1a_2 + b_1b_2)$;

13.11 (a) $n + m = p$, closure; $n + 0 = n$, identity;
$n + (-n) = (-n) + n = 0$, inverse;
$(n + m) + p = n + (m + p)$, associativity. For integer multiplication, 1
is a potential identity but, for $n \neq \pm 1$ the inverse $1/n$ is not in the set of
integers. With subtraction of integers, associativity fails, since
$a - (b - c) \neq (a - b) - c$,   (b) $ab = c$, closure; $a(1/a) = 1$,
inverse; $a1 = 1a = a$, identity; $a(bc) = (ab)c$, associativity;

13.12 (a) $x\mathbf{E}_2 = x\mathbf{E}_2\mathbf{E}_2 = x\mathbf{E}_2 = x$, (similarly $\mathbf{E}_2x = x$),

(b) $\mathbf{x}\mathbf{x}^{-1} = x\mathbf{E}_2 \cdot (1/x)\mathbf{E}_2 = 1 \cdot \mathbf{E}_2 = e$, (c) $\mathbf{xy} = \begin{pmatrix} xy & 0 \\ 0 & xy \end{pmatrix}$, which is a member of the set;

13.13 (b) Notice that $\mathbf{Z}_i^T = x_i\begin{pmatrix} 1 & 0 \\ 0 & 1 \end{pmatrix} + y_i\begin{pmatrix} 0 & -1 \\ 1 & 0 \end{pmatrix} = x_i\mathbf{E}_2 - y_i\mathbf{J}$;

13.14 (a) $\mathbf{D}(E) = \begin{pmatrix} 1 & 0 & 0 \\ 0 & 1 & 0 \\ 0 & 0 & 1 \end{pmatrix}$, $\mathbf{D}(C_2(z)) = \begin{pmatrix} 1 & 0 & 0 \\ 0 & 0 & 1 \\ 0 & 1 & 0 \end{pmatrix}$,

$\mathbf{D}(\sigma(yz)) = \begin{pmatrix} 1 & 0 & 0 \\ 0 & 1 & 0 \\ 0 & 0 & 1 \end{pmatrix}$, $\mathbf{D}(\sigma(xz)) = \begin{pmatrix} 1 & 0 & 0 \\ 0 & 0 & 1 \\ 0 & 1 & 0 \end{pmatrix}$,

(b) $\mathbf{D}(\sigma(yz)) = \begin{pmatrix} \mathbf{X}(\sigma(yz)) & \mathbf{O}_3 & \mathbf{O}_3 \\ \mathbf{O}_3 & \mathbf{X}(\sigma(yz)) & \mathbf{O}_3 \\ \mathbf{O}_3 & \mathbf{O}_3 & \mathbf{X}(\sigma(yz)) \end{pmatrix}$, where

$\mathbf{X}(\sigma(yz)) = \begin{pmatrix} -1 & 0 & 0 \\ 0 & 1 & 0 \\ 0 & 0 & 1 \end{pmatrix}$,

$\mathbf{D}(\sigma(xz)) = \begin{pmatrix} \mathbf{X}(\sigma(xz)) & \mathbf{O}_3 & \mathbf{O}_3 \\ \mathbf{O}_3 & \mathbf{O}_3 & \mathbf{X}(\sigma(xz)) \\ \mathbf{O}_3 & \mathbf{X}(\sigma(xz)) & \mathbf{O}_3 \end{pmatrix}$, where

$$\mathbf{X}(\sigma(xz)) = \begin{pmatrix} 1 & 0 & 0 \\ 0 & -1 & 0 \\ 0 & 0 & 1 \end{pmatrix}.$$

14   Determinants – functions revisited and a new notation

14.1 -12, -12, -6, -1;

14.2 (a) Step 1: remove 2 from each row (or column), step 2: denoting the jth column by c_j, $c_1 \to c_1 - 2c_3$ and $c_2 \to c_2 - c_3$; (b) see 14.1;

14.3 Denoting the ith row by r_i, let $r_1 \to r_1 - r_2$ and $r_2 \to r_2 - r_3$;

14.4 $r_1 \to r_1 + r_2 + r_3 + r_4 + r_5$, then $c_5 \to c_5 - c_1$, $c_4 \to c_4 - c_1$ and so on, to give

$$\begin{vmatrix} x + 4y & 0 & 0 & 0 & 0 \\ y & x - y & 0 & 0 & 0 \\ y & 0 & x - y & 0 & 0 \\ y & 0 & 0 & x - y & 0 \\ y & 0 & 0 & 0 & x - y \end{vmatrix} = (x - y)^4 (4y + x),$$

14.5 $\begin{vmatrix} 0 & \cos x \\ x & x^2 \end{vmatrix} + \begin{vmatrix} 1 & \sin x \\ 1 & 2x \end{vmatrix} = -x \cos x + 2x - \sin x$;

14.6 (a) $\cos 2\theta$, (b) $-2 \sin 2\theta$, $-4 \cos 2\theta$, (c) turning points when $\sin 2\theta = 0 \Rightarrow \theta = n\pi/2$, yielding maxima and minima when n is even or odd, respectively;

14.7 (a) $A = 14$, $B = \begin{pmatrix} 9 & 3 & 6 \\ 3 & 1 & 2 \\ 6 & 2 & 4 \end{pmatrix}$, (b) $\lambda = 14$; $\lambda = 0, 0, 14$;

14.8 (c) $\lambda_1 = -1$ or 2,
(d) $b = -1$, $c = -2$ ($\lambda_1 = -1$), or $b = 2$, $c = 1$ ($\lambda_1 = 2$),
(e) $\lambda_2 = -1$ and $\lambda_3 = 2$ ($\lambda_1 = -1$), or $\lambda_2 = -1$ and $\lambda_3 = -1$ ($\lambda_1 = 2$);

14.9 (a) $\lambda = -1$, (b) $\lambda = 3$, (c) $\mathbf{F} = (1/\sqrt{2})\begin{pmatrix} 1 \\ 1 \end{pmatrix}$, (d) $\mathbf{F}^T\mathbf{G} = 0$;

14.10 $A^{11} = \begin{vmatrix} a_{22} & a_{23} \\ a_{32} & a_{33} \end{vmatrix}$, $A^{31} = \begin{vmatrix} a_{12} & a_{13} \\ a_{22} & a_{23} \end{vmatrix}$,

$A^{12} = -\begin{vmatrix} a_{21} & a_{23} \\ a_{31} & a_{33} \end{vmatrix}$, $A^{22} = \begin{vmatrix} a_{11} & a_{13} \\ a_{31} & a_{33} \end{vmatrix}$,

$A^{32} = -\begin{vmatrix} a_{11} & a_{13} \\ a_{21} & a_{23} \end{vmatrix}$, $A^{13} = \begin{vmatrix} a_{21} & a_{22} \\ a_{31} & a_{32} \end{vmatrix}$,

$A^{23} = -\begin{vmatrix} a_{11} & a_{12} \\ a_{31} & a_{32} \end{vmatrix}$, $A^{33} = \begin{vmatrix} a_{11} & a_{12} \\ a_{21} & a_{22} \end{vmatrix}$;

14.11 (a) $A^{11} = 12$, $A^{22} = 2$, $A^{12} = -8$, $A^{21} = -6$, (b) $A^{12} = -y(x - y)^3$;

14.12 Notice that in each case the sum of products of elements and cofactors generates a determinant in which two rows (columns) are identical;

14.14 $A^{-1} = \frac{1}{12} \begin{pmatrix} -12 & 6 & -12 \\ 8 & -2 & 4 \\ -6 & 3 & 0 \end{pmatrix}$;

14.15 (a)(i) $\frac{1}{10} \begin{pmatrix} 5 & 3 & -1 \\ -5 & 1 & 3 \\ -5 & -5 & 5 \end{pmatrix}$, (ii) $\frac{1}{10} \begin{pmatrix} -4 & 10 & -8 \\ -15 & 5 & -15 \\ -8 & 0 & -6 \end{pmatrix}$,

(iii) $\begin{pmatrix} \cos\theta & \sin\theta & 0 \\ -\sin\theta & \cos\theta & 0 \\ 0 & 0 & 1 \end{pmatrix}$, (b)(i) $x^T = (\,2 \quad -1 \quad -3\,)$,

(b)(ii) $x^T = (\,0 \quad -2 \quad -1\,)$,

(b)(iii) $x^T = (\,x'\cos\theta + y'\sin\theta \quad -x'\sin\theta + y'\cos\theta \quad z'\,)$;

14.16 Non-trivial solution if $\begin{vmatrix} k & -4 & 1 \\ 2 & -2 & -k \\ 1 & 2 & -2 \end{vmatrix} = 0 \Rightarrow k = 1, -5$;

$k = 1 : x = 2/\sqrt{5},\ y = 1/\sqrt{5},\ z = 2/\sqrt{5}$,

$k = -5 : x = -2/\sqrt{17},\ y = 3/\sqrt{17},\ z = 2/\sqrt{17}$;

15   Vectors – a formalism for directional properties

15.1 (a) $\hat{i} + \hat{j} + 3\hat{k}$, (b) $3\hat{i} + \hat{j} - 5\hat{k}$, (c) $\hat{i} + 2\hat{j} + 10\hat{k}$, (d) $\sqrt{59}$,
   (e) $(2\hat{i} + \hat{j} - \hat{k})/\sqrt{6}$, (f) $\sqrt{6} + \sqrt{17}$, (g) $(\hat{i} + 7\hat{k})/2$;

15.2 (a) 23, (b) 34, (c) 12, (d) 58, (e) 1.09^c, (f) -12;

15.3 (a) $h_1 = R(\hat{i} + \hat{j} + \hat{k})/\sqrt{3}$, $h_2 = R(-\hat{i} - \hat{j} + \hat{k})/\sqrt{3}$,
   $h_3 = R(-\hat{i} + \hat{j} - \hat{k})/\sqrt{3}$, $h_4 = R(\hat{i} - \hat{j} - \hat{k})/\sqrt{3}$, (b) $109.5°$,
   (c) $H_1 - H_2 = 2R\sqrt{2/3}$;

15.4 (a) $C_1 - Cl = 179.5$ pm, $C_1 - C_2 = 152.3$ pm, and all the bonded C–H
   distances are 111.3 pm, (b) $103.50°$, $118.55°$;

15.5 (a) In each case $\theta = 0$, (b) in each case $\theta = \pi/2$;

15.6 (a) $-5\hat{i} + 11\hat{j} + \hat{k}$, (b) $5\hat{i} - 11\hat{j} - \hat{k}$, (c) $2\sqrt{14}$, (d) $4\hat{i} - 6\hat{j} + 2\hat{k}$,
   (e) -14;

15.7 Area is twice the area of the triangle OPQ, that is
   $2 \times |a|h/2 = |a||b|\sin\theta = |a \times b|$;

15.9 (a) Expand the 2×2 determinant and rearrange terms, (b) $y = 3x - 7$;

15.10 (b) Observe that $r \cdot n = Ax + By + Cz$, and $r_i \cdot n = Ax_i + By_i + Cz_i$;

15.11 (a) $63.43°(1.107^c)$, $54.74°(0.955^c)$, and $39.23°(0.685^c)$,
   (b) $V = 6$ units of volume,
   (c) $x_1 = 1/3$, $y_1 = -1/6$, $z_1 = -1/6$; $x_2 = 0$, $y_2 = 1/2$, $z_2 = -1/2$;
   $x_3 = 0$, $y_3 = 0$, $z_3 = 1$, (d) b_1 is in the direction of a_1, and orthogonal
   to a_2 and a_3, that is orthogonal to the plane containing a_2 and a_3,
   (e) $b_2 = (a_3 \times a_1)/V$, $b_3 = (a_1 \times a_2)/V$,

   (g) $X = \frac{1}{6} \begin{pmatrix} 2 & 3 & 0 \\ -1 & -3 & 0 \\ -1 & 0 & 6 \end{pmatrix}$, $Y = \begin{pmatrix} 3 & 1 & 1 \\ 0 & 2 & 1 \\ 0 & 0 & 1 \end{pmatrix}$,

(i) $\mathbf{Y}^{-1} = \frac{1}{6}\begin{pmatrix} 2 & -1 & -1 \\ 0 & 3 & -3 \\ 0 & 0 & 6 \end{pmatrix}$;

15.12 (b)(i) $\partial z/\partial r = 0$ either when $\cos\theta = \pi/2$ or when $\tan\theta = 1/2$ i.e., when $\theta + n\pi = \arctan\left(\frac{1}{2}\right)$. Using the result (see Problem 2.11(d)) $\tan(A + B) = \tan A$ if $B = n\pi$, yields $\theta = 26.56°, -153.44°$ (206.56°), (ii) the turning points of $\partial z/\partial r$ occur when $\theta = -31.7°, -121.7°, -211.7°, -301.7°, \ldots$ and second derivatives are positive when $\theta = -121.7°, -301.7°$ (or 238.3° and 58.3°, respectively), (c) $\hat{r}$ and $\hat{k}$ are in, and orthogonal to, the xy plane, respectively; $\hat{\theta} = \hat{k} \times \hat{r}$, which is orthogonal to $\hat{k}$, therefore lies in the xy plane, is orthogonal to $\hat{r}$, and has the form $\hat{\theta} = -\hat{i}\sin\theta + \hat{j}\cos\theta$;

15.13 $\partial r/\partial x = \cos\theta = x/\sqrt{x^2 + y^2}$,
$\partial r/\partial y = \sin\theta = y/\sqrt{x^2 + y^2}$,
$\partial\theta/\partial x = -(1/r)\sin\theta = -y/(x^2 + y^2)$,
$\partial\theta/\partial y = (1/r)\cos\theta = x/(x^2 + y^2)$;

15.14 (a) $\mathbf{J} = \begin{pmatrix} \frac{\partial x}{\partial r} & \frac{\partial x}{\partial\theta} & \frac{\partial x}{\partial\phi} \\ \frac{\partial y}{\partial r} & \frac{\partial y}{\partial\theta} & \frac{\partial y}{\partial\phi} \\ \frac{\partial z}{\partial r} & \frac{\partial z}{\partial\theta} & \frac{\partial z}{\partial\phi} \end{pmatrix}$

$= \begin{pmatrix} \sin\theta\cos\phi & r\cos\theta\cos\phi & -r\sin\theta\sin\phi \\ \sin\theta\sin\phi & r\cos\theta\sin\phi & r\sin\theta\cos\phi \\ \cos\theta & -r\sin\theta & 0 \end{pmatrix}$, (c) $5a_0$.

16   The eigenvalue problem – an important link between theory and experiment

16.1 (a) $\lambda = 1$, $\begin{pmatrix} 1/\sqrt{2} \\ -1/\sqrt{2} \end{pmatrix}$, $\lambda = 6$, $\begin{pmatrix} 4/\sqrt{17} \\ 1/\sqrt{17} \end{pmatrix}$, (b) $\lambda = 1$, $\begin{pmatrix} 1 \\ 0 \end{pmatrix}$,

(c) expand $\det(\mathbf{R} - \mathbf{E}\lambda)$ from the second row to yield

$\lambda = 4$, $\begin{pmatrix} 1/\sqrt{14} \\ -2/\sqrt{14} \\ -3/\sqrt{14} \end{pmatrix}$, $\lambda = -2$, $\begin{pmatrix} -1/\sqrt{10} \\ 0 \\ 3/\sqrt{10} \end{pmatrix}$,

$\lambda = 8$, $\begin{pmatrix} 1/\sqrt{2} \\ 0 \\ -1/\sqrt{2} \end{pmatrix}$,

(d) $\lambda = 3$, $\begin{pmatrix} 1 \\ 0 \\ 0 \end{pmatrix}$, $\lambda = 1$, $\begin{pmatrix} 0 \\ 1 \\ 0 \end{pmatrix}$,

$\lambda = 2$, $\begin{pmatrix} 0 \\ 0 \\ 1 \end{pmatrix}$;

16.2 $\lambda = (\alpha - \beta)$ twice, $\dfrac{1}{\sqrt{2}}\begin{pmatrix} 0 \\ 1 \\ -1 \end{pmatrix}$ and $\dfrac{1}{\sqrt{2}}\begin{pmatrix} 1 \\ 0 \\ -1 \end{pmatrix}$

(non-orthogonal eigenvectors),

$$\dfrac{1}{\sqrt{2}}\begin{pmatrix} 0 \\ 1 \\ -1 \end{pmatrix} \text{ and } \dfrac{1}{\sqrt{6}}\begin{pmatrix} -2 \\ 1 \\ 1 \end{pmatrix}$$

(orthogonal eigenvectors),

$$\lambda = (\alpha + 2\beta),\ \dfrac{1}{\sqrt{3}}\begin{pmatrix} 1 \\ 1 \\ 1 \end{pmatrix};$$

16.4 (a) $\hat{i}'$ could be in the first or fourth quadrants but, since $\hat{i}' \cdot \hat{j} = 1/\sqrt{3}$, it is in the fourth quadrant with $\theta_y = 54.7°$,
(b) $\hat{j}' \cdot \hat{i} = 1/\sqrt{3}$, $\hat{j}' \cdot \hat{j} = -\sqrt{2/3}$, yielding $\theta_x = 54.7°$, $\theta_y = 144.7°$, with $\hat{j}'$ in the fourth quadrant,
(c) changing the overall sign of the first eigenvector representatives yields $\hat{i}'$ in the second quadrant ($\theta_x = 144.7°$, $\theta_y = 54.7°$) and $\hat{j}'$ in the third quadrant ($\theta_x = 125.3°$, $\theta_y = 144.7°$); changing the overall sign of the second eigenvector representative yields $\hat{i}'$ in the fourth quadrant ($\theta_x = 35.3°$, $\theta_y = 125.3°$) and $\hat{j}'$ in the first quadrant ($\theta_x = 54.7°$, $\theta_y = 35.3°$); changing the overall sign of both eigenvector representative yields $\hat{i}'$ in the third quadrant ($\theta_x = 144.7°$, $\theta_y = 125.3°$) and $\hat{j}'$ in the second quadrant ($\theta_x = 125.3°$, $\theta_y = 35.3°$).

17   Curve fitting – vectors revisited

17.1 (a) $a_1 = (2x_1 - x_2 - x_3)/2$, $a_2 = (x_2 + x_3)/2$,
(b) $a_1 = 1/3, a_2 = 1/3$.

Index